NATIONAL CONTRACTOR'S EXAM STUDY GUIDE

About the Author

R. Dodge Woodson is a seasoned builder of as many as 60 single family homes a year, a master re-modeler, a master plumber and a master gasfitter with over 30 years of experience. Woodson opened his own business in 1979 and is the owner of The Masters Group, Inc., in Brunswick, Maine. In addition to owning and operating his contracting business, R. Dodge Woodson has taught both code and apprentice classes in the technical college system in Maine. Well known as a prolific author of many McGraw-Hill titles, Woodson's reputation and experience come together to offer readers a real-life view of professional preparation for passing the licensing exam for a trade license.

NATIONAL CONTRACTOR'S EXAM STUDY GUIDE

R. Dodge Woodson
Jane L. Roy, Editor

New York Chicago San Francisco Lisbon London Madrid
Mexico City Milan New Delhi San Juan Seoul
Singapore Sydney Toronto

The **McGraw·Hill** Companies

CIP Data is on file with the Library of Congress.

31143007695613
690.076 Woodson
Woodson, R. Dodge (Roger
Dodge), 1955-
National contractor's
exam study guide

McGraw-Hill books are available at special quantity discounts to use as premiums and sales promotions, or for use in corporate training programs. For more information, please write to the Director of Special Sales, Professional Publishing, McGraw-Hill, Two Penn Plaza, New York, NY 10121-2298. Or contact your local bookstore.

1 2 3 4 5 6 7 8 9 0 DOC/DOC 0 1 3 2 1 0 9 8 7

ISBN-13: 978-0-07-148907-2
ISBN-10: 0-07-148907-X

Sponsoring Editor	**Proofreader**
Cary Sullivan	Leona Woodson
Editing Supervisor	**Production Supervisor**
David E. Fogarty	Pamela A. Pelton
Project Manager	**Composition**
Jacquie Wallace	Lone Wolf Enterprises, Ltd.
Copy Editor	**Art Director, Cover**
Wendy Lochner	Jeff Weeks

This book is printed on acid-free paper.

This book is dedicated to
Leona, for making me a happy man.

Contents

Preface

Passing the licensing exam for a contractor's license is a major step toward a potentially lucrative career. A licensed contractor who goes into business can make a lot of money. Even working for an employer can give a builder a better-than-average wage. There is money in building, and it's usually not too difficult to find work. But, one of the first steps is passing the licensing exam. Many potential contractors fail their licensing exams a time or two before they pass them. But, this book can help you be one of the people in the exam room who walks away with a passing grade and a trade license.

I've been in the trades for over 25 years. It was 1979 when I went into business for myself. I am a Class A builder, a master plumber and a master gasfitter. During my career I have taught as adjunct faculty for Central Maine Technical College. One of the classes I taught was a code preparation class. This course was designed to help people pass their licensing exams. While teaching that class, I learned a lot about what potential licensees have trouble with in studying for a licensing exam. While I can't look you in the eye and teach you one on one, I can help you with this book. My past students have enjoyed a high success ratio in passing their exams. There is no reason why you should be left behind, now that this book is available to you.

Why do you need a study guide? The code can be difficult to understand and to interpret. Wording on exams can be very tricky. While you might be well versed in the code for field work, it would not be unusual for you to have difficulty passing an exam on the code regulations. Some people simply have trouble taking tests. The sample tests here will give you a feel for what to expect from your licensing exam. After taking these tests and passing them, you will have more confidence on the day when you take your real exam. There will be no need to freeze up when your exam is placed in front of you. Since you will have tested your knowledge here and studied your local code book, you will be well prepared for the licensing exam.

There is more here than simply a collection of test questions and answers. Take a moment to thumb through the pages. Notice all of the illustrations. They aren't pretty pictures used to make a bigger book. The illustrations are a type of nuts-and-bolts help for your study needs and your field work.

What is a trade license worth to you? It should be worth a lot. Fortunately, you don't have to pay a lot for it. All you need is this book, your local code book, some good study habits, and funds for your licensing exam. Investing in yourself and your license may well prove to be one of the best investments you will ever make. Go for it.

R. Dodge Woodson

Acknowledgments

I would like to thank Jane Roy for her help in putting this project together and The International Code Conference for allowing me to work with their code materials in the creation of this study guide.

As always, special thanks go out to Ginny and Jacquie and the rest of the Lone Wolf team for their professional expertise in the production process of this project.

Introduction

Techniques for Studying and Test-Taking

PREPARING FOR THE EXAM

1. Make a study schedule. Assign yourself a period of time each day to devote to preparation for your exam. A regular time is best, but the important thing is daily study.

2. Study alone. You will concentrate better when you work by yourself. Keep a list of questions you find puzzling and points you are unsure of to talk over with a friend who is preparing for the same exam. Plan to exchange ideas at a joint review session just before the test.

3. Eliminate distractions. Choose a quiet, well-lit spot as far as possible from telephone, television, and family activities. Try to arrange not to be interrupted.

4. Begin at the beginning. Read. Underline points that you consider significant. Make marginal notes. Flag the pages that you think are especially important with little Post-it™ Notes.

5. Concentrate on the information and instruction chapters. Study the Code Definitions, the Glossary of air-conditioning and refrigeration terms, and the Decoding Words of equipment and usage. Learn the language of the field. Focus on the technique of eliminating wrong answers. This information is important to answering all multiple-choice questions.

6. Answer the practice questions chapter by chapter. Take note of your weaknesses; use all available textbooks to brush up.

7. Try the previous exams if available. When you believe that you are well prepared, move on to these exams. If possible, answer an entire exam in one sitting. If you must divide your time, divide it into no more than two sessions per exam.

- When you do take the practice exams, treat them with respect. Consider each as a dress rehearsal for the real thing. Time yourself accurately, and do not peek at the correct answers.

- Remember, you are taking these for practice; they will not be scored; they do not count. So learn from them.

IMPORTANT: Do not memorize questions and answers. Any question that has been released will not be used again. You may run into questions that are very similar, but you will not be tested with the original ones. The included questions will give you good practice, but they will not be the same as any of the questions on your exam.

HOW TO TAKE AN EXAM

Get to the examination room about 10 minutes ahead of time. You'll get a better start when you are accustomed to the room. If the room is too cold, too warm, or not well ventilated, call these conditions to the attention of the person in charge.

Make sure that you read the instructions carefully. In many cases, test takers lose points because they misread some important part of the directions. (An example would be reading the incorrect choice instead of the correct choice.)

Don't be afraid to guess. The best policy is, of course, to pace yourself so that you can read and consider each question. Sometimes this does not work. Most exam scores are based only on the number of questions answered correctly. This means that a wild guess is better than a blank space. There is no penalty for a wrong answer, and you just might guess right. If you see that time is about to run out, mark all the remaining spaces with the same answer. According to the law of averages, some will be right.

However, you bought this book for practice in answering questions. Part of your preparation is learning to pace yourself so that you need not answer randomly at the end. Far better than a wild guess is an educated guess. You make this kind of guess not when you are pressed for time but are not sure of the correct answer. Usually, one or two of the choices are obviously wrong. Eliminate the obviously wrong answers and try to reason among those remaining. Then, if necessary, guess from the smaller field. The odds of choosing a right answer increase if you guess from a field of two instead of from a field of four. When you make an educated guess or a wild guess in the course of the exam, you might want to make a note next to the question number in the test booklet. Then, if there is time, you can go back for a second look.

Reason your way through multiple-choice questions very carefully and methodically.

MULTIPLE-CHOICE TEST-TAKING TIPS

Here are a few examples that we can "walk through" together:

1.   On the job, your supervisor gives you a hurried set of directions. As you start your assigned task, you realize you are not quite clear on the directions given to you. The best action to take would be to:

   (a) continue with your work, hoping to remember the directions

   (b) ask a co-worker in a similar position what he or she would do

(c) ask your supervisor to repeat or clarify certain directions

(d) go on to another assignment

In this question you are given four possible answers to the problem described. Though the four choices are all possible actions, it is up to you to choose the best course of action in this particular situation.

Choice (a) will likely lead to a poor result; given that you do not recall or understand the directions, you would not be able to perform the assigned task properly. Keep choice (a) in the back of your mind until you have examined the other alternatives. It could be the best of the four choices given.

Choice (b) is also a possible course of action, but is it the best? Consider that the co-worker you consult has not heard the directions. How could he or she know? Perhaps his or her degree of incompetence is greater than yours in this area. Of choices (a) and (b), the better of the two is still choice (a).

Choice (c) is an acceptable course of action. Your supervisor will welcome your questions and will not lose respect for you. At this point, you should hold choice (c) as the best answer and eliminate choice (a).

The course of action in choice (d) is decidedly incorrect because the job at hand would not be completed. Going on to something else does not clear up the problem; it simply postpones your having to make a necessary decision.

After careful consideration of all choices given, choice (c) stands out as the best possible course of action. You should select choice (c) as your answer.

Every question is written about a fact or an accepted concept. The question above indicates the concept that, in general, most supervisory personnel appreciate subordinates questioning directions that may not have been fully understood. This type of clarification precludes subsequent errors. On the other hand, many subordinates are reluctant to ask questions for fear that their lack of understanding will detract from their supervisor's evaluation of their abilities.

The supervisor, therefore, has the responsibility of issuing orders and directions in such a way that subordinates will not be discouraged from asking questions. This is the concept on which the sample question was based.

Of course, if you were familiar with this concept, you would have no trouble answering the question. However, if you were not familiar with it, the method outlined here of eliminating incorrect choices and selecting the correct one should prove successful for you.

We have now seen how important it is to identify the concept and the key phrase of the question. Equally or perhaps even more important is identifying and analyzing the keyword or the qualifying word in a question. This word is usually an adjective or adverb. Some of the most common key words are:

most	least	best	highest		
lowest	always	never	sometimes		
most likely	greatest	smallest	tallest		
average	easiest	most nearly	maximum		
minimum	only	chiefly	mainly	but	or

Identifying these keywords is usually half the battle in understanding and, consequently, answering all types of exam questions.

Identifying these keywords is usually half the battle in understanding and, consequently, answering all types of exam questions.

Now we will use the elimination method on some additional questions.

2. On the first day you report for work after being appointed as an AC mechanic's helper, you are assigned to routine duties that seem to you to be very petty in scope. You should:

   (a) perform your assignment perfunctorily while conserving your energies for more important work in the future

   (b) explain to your superior that you are capable of greater responsibility

   (c) consider these duties an opportunity to become thoroughly familiar with the workplace

   (d) try to get someone to take care of your assignment until you have become thoroughly acquainted with your new associates

Once again we are confronted with four possible answers from which we are to select the best one.

Choice (a) will not lead to getting your assigned work done in the best possible manner in the shortest possible time. This would be your responsibility as a newly appointed AC mechanic's helper, and the likelihood of getting to do more important work in the future following the approach stated in this choice is remote. However, since this is only choice (a), we must hold it aside because it may turn out to be the best of the four choices given.

Choice (b) is better than choice (a) because your superior may not be familiar with your capabilities at this point. We therefore should drop choice (a) and retain choice (b) because, once again, it may be the best of the four choices.

The question clearly states that you are newly appointed. Therefore, would it not be wise to perform whatever duties you are assigned in the best possible manner? In this way, you would not only use the opportunity to become acquainted with procedures but also to demonstrate your abilities.

Choice (c) contains a course of action that will benefit you and the location in which you are working because it will get needed work done. At this point, we drop choice (b) and retain choice (c) because it is by far the better of the two.

The course of action in choice (d) is not likely to get the assignment completed, and it will not enhance your image to your fellow AC mechanic's helpers.

Choice (c), when compared to choice (d), is far better and therefore should be selected as the best choice.

Now let us take a question that appeared on a police-officer examination:

3. An off-duty police officer in civilian clothes riding in the rear of a bus notices two teenage boys tampering with the rear emergency door. The most appropriate action for the officer to take is to:

   (a) tell the boys to discontinue their tampering, pointing out the dangers to life that their actions may create

   (b) report the boys' actions to the bus operator and let the bus operator take whatever action is deemed best

(c) signal the bus operator to stop, show the boys the officer's badge, and then order them off the bus

(d) show the boys the officer's badge, order them to stop their actions, and take down their names and addresses

Before considering the answers to this question, we must accept that it is a well-known fact that a police officer is always on duty to uphold the law even though he or she may be technically off duty.

In choice (a), the course of action taken by the police officer will probably serve to educate the boys and get them to stop their unlawful activity. Since this is only the first choice, we will hold it aside.

In choice (b), we must realize that the authority of the bus operator in this instance is limited. He can ask the boys to stop tampering with the door, but that is all. The police officer can go beyond that point. Therefore, we drop choice (b) and continue to hold choice (a).

Choice (c) as a course of action will not have a lasting effect. What is to stop the boys from boarding the next bus and continuing their unlawful action? We therefore drop choice (c) and continue to hold choice (a).

Choice (d) may have some beneficial effect, but it would not deter the boys from continuing their actions in the future.

When we compare choice (a) with choice (d), we find that choice (a) is the better one overall, and therefore it is the correct answer.

The next question illustrates a type of question that has gained popularity in recent examinations and that requires a two-step evaluation.

First, the reader must evaluate the condition in the question as being "desirable" or "undesirable." Once the determination has been made, we are then left with making a selection from two choices instead of the usual four.

4. A visitor to an office in a city agency tells one of the aides that he has an appointment with the supervisor, who is expected shortly. The visitor asks for permission to wait in the supervisor's private office, which is unoccupied at the moment. For the office aide to allow the visitor to do so would be:

    (a) desirable; the visitor would be less likely to disturb the other employees or to be disturbed by them

    (b) undesirable; it is not courteous to permit a visitor to be left alone in an office

    (c) desirable; the supervisor may wish to speak to the visitor in private

    (d) undesirable; the supervisor may have left confidential papers on the desk

First of all, we must evaluate the course of action on the part of the office aide of permitting the visitor to wait in the supervisor's office as being very undesirable. There is nothing said of the nature of the visit; it may be for a purpose that is not friendly or congenial. There may be papers on the supervisor's desk that he or she does not want the visitor to see or to have knowledge of. Therefore, at this point, we have to decide between choices (b) and (d).

This is definitely not a question of courtesy. Although all visitors should be treated with courtesy, permitting the visitor to wait in the supervisor's office is not the only possible act of courtesy. Another comfortable place could be found for the visitor to wait.

Choice (d) contains the exact reason for evaluating this course of action as being undesirable, and when we compare it with choice (b), choice (d) is far better.

TEST QUESTIONS AND ANSWERS

Keep in mind that tests or exams are also learning tools. They make you learn the assigned material so that you don't have to refer to sources other than your own brain's memory.

There are a number of types of questions utilized in everyday teaching and learning. The advantage of the multiple-choice type of question is that it makes you think and then utilize your reasoning power to arrive at an educated guess. That is, of course, if you didn't know the answer outright from previous experience or studying.

Next, there is the true-false type of question. It is easy to answer either true (T) or false (F), you are either right or wrong. You have a 50 percent chance of being right or wrong when you guess. One of the major reasons this type of test is used is its quick right or wrong answer. It makes you think fast and recall the material you recently read or studied and quickly focuses your learning on the necessary information.

Both types are easy to check and grade. They also make the instructor's role an easier one.

Another type of question and answer test is "fill-in the blank." This requires you to think in regards to the meaning of the sentence with the missing word. There are, however, many clues in the question or statement before you. This type of test can also be used with the blank filled in by four possible answers. This type then resembles the multiple-choice type of test and serves the same purpose.

There are certifying agencies organized and operating to aid you in obtaining the skills and knowledge to perform correctly in your chosen field. By requiring you to submit to a written exam on the material, you have an incentive to study hard and organize the material you have committed to memory. Passing the exam from one of the agencies makes it easier for them to hire you to do the work and know that you can do it with some degree of skill and perfection.

The inspector relies on you to do the assigned task properly and safely. The inspector has to be thoroughly familiar with all the contract documents, including the plans with all changes, specifications, and contract submittals such as shop drawings.

Inspectors have different responsibilities and authorities, depending on the organizational setup, and size and scope of the project. Each inspector should be clear on the answers to the many questions presented during a day in the field.

Inspectors have the task of examining a finished job and informing the specialist as to how his work meets or fails to meet the code requirements.

A STRATEGY FOR TEST DAY

On the exam day assigned to you, allow the test itself to be the main attraction of the day. Do not squeeze it in between other activities. Arrive rested, relaxed, and on time. In fact, plan to arrive a little bit early. Leave plenty of time for traffic tie-ups or other complications that might upset you and interfere with your test performance.

Here is a breakdown of what occurs on examination day and tips on starting off on the right foot and preparing to start your exam:

1.   In the test room the examiner will hand out forms for you to fill out and will give you the instructions that you must follow in taking the examination. Note that you must follow instructions exactly.

2. The examiner will tell you how to fill in the blanks on the forms.

3. Exam time limits and timing signals will be explained.

4. Be sure to ask questions if you do not understand any of the examiner's instructions. You need to be sure that you know exactly what to do.

5. Fill in the grids on the forms carefully and accurately. Filling in the wrong blank may lead to loss of veterans' credits to which you may be entitled or to an incorrect address for your test results.

6. Do not begin the exam until you are told to begin.

7. Stop as soon as the examiner tells you to stop.

8. Do not turn pages until you are told to do so.

9. Do not go back to parts you have already completed.

10. Any infraction of the rules is considered cheating. If you cheat, your test paper will not be scored, and you will not be eligible for appointment.

11. Once the signal has been given and you begin the exam, read every word of every question.

12. Be alert for exclusionary words that might affect your answer: words such as "not" "most," and "least."

MARKING YOUR ANSWERS

Read all the choices before you mark your answer. It is statistically true that most errors are made when the last choice is the correct answer. Too many people mark the first answer that seems correct without reading through all the choices to find out which answer is best.

Be sure to read the suggestions below now and review them before you take the actual exam. Once you are familiar with the suggestions, you will feel more comfortable with the exam itself and find them useful when you are marking your answer choices.

1. Mark your answers by completely blackening the answer space of your choice.

2. Mark only ONE answer for each question, even if you think that more than one answer is correct. You must choose only one. If you mark more than one answer, the scoring machine will consider you wrong even if one of your answers is correct.

3. If you change your mind, erase completely. Leave no doubt as to which answer you have chosen.

4. If you do any figuring on the test booklet or on scratch paper, be sure to mark your answer on the answer sheet.

5. Check often to be sure that the question number matches the answer space number and that you have not skipped a space by mistake. If you do skip a space, you must erase all the answers after the skip and answer all the questions again in the right places.

6. Answer every question in order, but do not spend too much time on any one question.

If a question seems to be "impossible," do not take it as a personal challenge. Guess and move on. Remember that your task is to answer correctly as many questions as possible. You must apportion your time so as to give yourself a fair chance to read and answer all the questions. If you guess at an answer, mark the question in the test booklet so that you can find it again easily if time allows.

7.   Guess intelligently if you can. If you do not know the answer to a question, eliminate the answers that you know are wrong and guess from among the remaining choices. If you have no idea whatsoever of the answer to a question, guess anyway. Choose an answer other than the first.

      The first choice is generally the correct answer less often than the other choices. If your answer is a guess, either an educated guess or a wild one, mark the question in the question booklet so that you can give it a second try if time permits.

8.   If you happen to finish before time is up, check to be sure that each question is answered in the right space and that there is only one answer for each question. Return to the difficult questions that you marked in the booklet and try them again. There is no bonus for finishing early so use all your time to perfect your exam paper.

With the combination of techniques for studying and test taking as well as the self-instructional course and sample examinations in this book, you are given the tools you need to score high on your exam.

Chapter 1
ADMINISTRATION

National Contractor's Exam Study Guide

As a contractor, you must be familiar with the administrative aspects of the building code. You may be asking yourself why. I'll tell you. As a contractor, it is important for you to know how the building code is enforced. The administrative chapter tells you whom you should contact for questions or what information you need to include when obtaining a permit for the work you are going to perform. Without this knowledge, you may apply for a permit without realizing that you need to send in copies of your construction documents, or you may not know who has the right to modify those plans. Make sure that you have read and understood the administrative chapter of the code book; it is there to help you understand the steps to take, the documentation to include, and who to contact in case of questions regarding the code. Let's move to the Chapter 1 test, which includes both multiple-choice and true-false questions.

MULTIPLE-CHOICE QUESTIONS

1. Detached one- and two-family dwellings must comply with:
   a. The International Fire Code
   b. The International Building Code
   c. The International Residential Code
   d. all of the above

2. The purpose of the IBC code is to make sure the minimum requirements to protect public health, safety, and general _____ are met.
   a. sanitation   b. repair
   c. systems   d. welfare

3. The ICC Electrical Code applies to the installation of:
   a. gas piping   b. mechanical systems
   c. electrical systems   d. ventilation

4. Installation and repair of a water or sewage system must be done in compliance with this code:
   a. The International Plumbing Code
   b. The International Fuel Gas Code
   c. The International Private Sewage Disposal Code
   d. The International Building Code

5. The International Property Maintenance Code applies to which of the following:

   a. equipment and facilities

   b. energy efficiency

   c. residential and commercial gas appliances

   d. all of the above

6. The building official has the authority to:

   a. review construction documents

   b. issue notices or orders

   c. grant modifications

   d. all of the above

7. Any owner can obtain a building permit after the following step is taken:

   a. calling the building official   b. calling the local code officer

   c. writing a letter   d. filling out an application

8. Permits are not required for which of these?

   a. fences not over 6 feet high   b. moving a building

   c. replacing electricity   d. altering a building

9. Emergency repairs must be reported to the building official within:

   a. 3 days   b. 1 day

   c. 2 days   d. the same day

10. Applications are not required for ordinary repairs, which do not include:

    a. replacement of lamps   b. repairing plumbing leaks

    c. removal of structural beams   d. replacement of electrical outlets

11. Permits are not required of the following:

    a. homeowners   b. electricians

    c. plumbers   d. public-service agencies

12. An application for a permit is considered to be abandoned after _____ days, unless an extension has been requested in writing.

    a. 90 days   b. 180 days

    c. 60 days   d. 30 days

13.   The building permit or copy must be kept in which of the following:

   a.   a safe   b.   the code office

   c.   on site   d.   in a deposit box

14.   Construction documents must be prepared by:

   a.   a registered design professional

   b.   an artist

   c.   the building official

   d.   a construction worker

15.   The construction documents must show, in detail, the location, construction, size, and character of:

   a.   exits   b.   bedrooms

   c.   living space   c.   fireplaces

16.   An approved construction document must have writing or a stamp that states:

   a.   approved

   b.   reviewed for code compliance

   c.   evidence found to be satisfactory

   d.   acknowledged as approved

17.   _____ are defined as portions of the design that are not submitted at the time of the application.

   a.   late documents   b.   amended documents

   c.   phased documents   d.   deferred documents

18.   Any changes made during construction that are not in compliance with the approved construction documents must be submitted as:

   a.   amended documents   b.   phased documents

   c.   deferred documents   d.   late documents

19.   Construction work is required to be inspected and must be _____ and exposed for inspection purposes.

   a.   done   b.   satisfactory

   c.   accessible   d.   obstructed

20.   A certificate of occupancy must include the following:

a.   the design occupant load

b.   the building-permit number

c.   the name and address of the owner

d.   all of the above

21.   The building official is authorized to issue a temporary certificate of _____ before the completion of work.

a.   possession   b.   occupancy

c.   ownership   d.   property

22.   If anyone disagrees with a decision made by the building official, you must file with:

a.   the Board of Appeals   b.   the code official

c.   your local town official   d.   a judge

23.   Reports of inspections must be made in writing and _____:

a.   delivered by hand   b.   by telephone

c.   certified   d.   air-mailed

24.   Reports from any required testing must be kept by:

a.   your town office

b.   the building owner

c.   the agency that performed the testing

d.   the building official

25.   Any person who is granted an annual permit must keep a detailed record of:

a.   time sheets of workers

b.   alterations

c.   painting or other similar work

d.   equipment

TRUE-FALSE QUESTIONS

1. The provisions of the IBC code apply to painting, papering, tiling, or other similar finish work.

    True   False

2. The provisions of the International Private Sewage Disposal Code must be applied to the installation of plumbing systems.

    True   False

3. The building official is not liable for cost in any suit that occurs.

    True   False

4. Used equipment can be reused without approval from the building official.

    True   False

5. The building official has the authority to grant modifications at his or her discretion.

    True   False

6. Permits are not required for the construction of oil derricks.

    True   False

7. Replacement or relocation of a standpipe is considered to be a repair, and no permit is needed.

    True   False

8. The building official has the right to reject any permit application that does not conform to code.

    True   False

9. Construction documents must be dimensioned and drawn on correct material.

    True   False

10. The only time you need a site plan is for the construction of playgrounds.

    True   False

11. When approved, construction documents must always be stamped with the word "approved" in red ink.

    True   False

12. Electronic-media documents are allowed when approved by the building official.

    True   False

13. Your permit will not be valid until all fees have been paid to the building official.

    True   False

14. A one-time fee is charged for all the permits that you need.

    True   False

15. The building official is authorized to create a refund policy.

    True   False

16. The building official can make inspections without notice.

    True   False

17. Footing and foundation inspections must be approved before excavations are made.

    True   False

18. Protection of joints and penetrations in fire-resistance-rated assemblies must be concealed from view until inspected and approved.

    True   False

19. The building official will accept reports of approved inspection agencies if the agencies meet the requirements.

    True   False

20. The building inspector will call the owner to make an appointment for building inspection.

    True   False

21. A building owner can hook up fuel or power without a permit.

    True   False

22. A stop-work order is just a threat that the building official uses; it is never enforced.

    True   False

23.   The board of appeals cannot waive any requirement of the IBC code.
      True   False

24.   A permit valuation includes everything but labor.
      True   False

25.   An application for a permit must include the value of the work to be done.
      True   False

MULTIPLE-CHOICE ANSWER KEY

1. C	6. D	11. D	16. B	21. B
2. D	7. D	12. B	17. D	22. A
3. C	8. A	13. C	18. A	23. C
4. A	9. B	14. A	19. C	24. D
5. A	10. C	15. A	20. D	25. B

TRUE-FALSE ANSWER KEY

1. T	6. T	11. F	16. F	21. F
2. F	7. F	12. T	17. F	22. F
3. T	8. T	13. T	18. F	23. T
4. F	9. T	14. F	19. T	24. F
5. T	10. F	15. T	20. F	25. T

Chapter 2
DEFINITIONS

When some people hear the word "test," their palms get sweaty and nerves start popping. Tests are not given to upset people, but rather, to find out if you have read and actually understand the materials that you are reading. Chapter 2 is going to test you on the meaning of building definitions. Some of these questions appear to have more than one answer. If you've studied carefully, you should not have trouble selecting the correct choice. Most professional testing is timed, so try not to spend too much time on any one question. Sometimes it is best to move on to the next question and go back if time allows. The building code is full of definitions, some of which you may already know. For the definitions that are not familiar, you will need to take the time to examine and study. Everyone has his or her own style of learning; what works for one person will not work for another. Find your learning style and stick to that. If you are consistent in learning (not just memorizing) the definitions, it will work to your advantage when taking the real test. This test starts with a multiple-choice exam. Remember, there may be more than one answer. Choose the one that is most closely related to the code definition.

MULTIPLE-CHOICE QUESTIONS

1. What do the letters in AAC masonry stand for?

    a. autoclaved anchor concrete   b. anchor area construction

    c. area aerated concrete   d. aerated concrete

2. A continuous, unobstructed path is the definition for which of the following words?

    a. circulation path   b. accessible path

    c. accessible route   c. circulation route

3. _____ is finely divided solid material that, when dispersed in the air in proper proportions, could be ignited by a flame or spark.

    a. pixie dust   b. combustible dust

    c. combustible powder   d. talcum dust

4. Combustible liquid is defined as having a closed cup point at or above one of the following temperatures in degrees Fahrenheit:

    a. 100   b. 75

    c. 125   d. 110

5. Which of the following is capable of being readily ignited from common sources of heat or at a temperature of 600 degrees Fahrenheit?

    a. flammable solid   b. flammable liquid

    c. flammable material   d. none of these

!**Code**alert

Decorative materials are materials applied over the building's interior finish for decorative, acoustical, or other effects, including foam plastics and materials containing foam plastics. Floor coverings, ordinary window shades, and interior-finish materials 0.025 inch or less in thickness applied directly to and adhering tightly to a substrate are not considered decorative materials.

6. Hazardous materials are chemicals or substances that are physical or health hazards and are defined and classified by which of these codes?

    a. International Building Code

    b. International Gas and Fuel Code

    c. International Health and Safety Code

    d. International Fire Code

7. Materials that when mixed have the potential to react in a manner that generates heat, fumes, or gases that are hazardous to life or property are defined as one of the following:

    a. organic materials   b. incompatible materials

    c. health hazards   d. highly toxic

8. A device that is designed to discover the presence of fire and initiate action is defined as:

    a. automatic fire detector   b. fire-alarm system

    c. fire-protection system   d. fire-alarm signal

9. Bleachers are defined as one of the following:

    a. whitening products   b. cleaners

    c. tiered seating at stadiums   d. Clorox

10. The leading edge of treads of stairs and of landings at the top of stairway flights is defined as one of these:

    a. degree   b. rate

    c. route   d. nose

!Codealert

A dwelling unit is a single unit providing complete, independent living facilities for one or more persons, including permanent provisions for sleeping, eating, cooking, and sanitation.

11. Veneers that are secured with approved mechanical fasteners are called:

    a. adhered masonry veneer   b. exterior masonry veneer

    c. vinyl masonry veneer   d. anchored masonry veneer

12. An interlayment is a layer of felt or nonbituminous saturated felt that is no less than _____ inches wide.

    a. 20   b. 15

    c. 18   c. 13

13. Reroofing is also known as which of the following:

    a. roof covering   b. both c and d

    c. roof recovering   d. roof replacement

14. A scupper is what?

    a. an opening for water drainage from the roof

    b. supper in a cup

    c. an enclosed roof structure

    d. an opening for roof ventilation

15. An approved fabricator is defined as:

    a. a qualified liar

    b. a qualified seamstress

    c. a qualified person approved by the building official

    d. none of these

16. A metal rod, wire, or strap that secures masonry to its structural support is called:

    a. stabilization
    b. connector
    c. mooring
    d. anchor

17. This type of dimension is defined as having a specified dimension plus an allowance for the joints with which the units are to be laid.

    a. nominal
    b. actual
    c. specified
    c. both a and b

18. The commercial size in standard sawn and glued-laminated lumber grades is:

    a. particle board
    b. nominal size
    c. standard size
    d. oversize

19. Particle board is a generic term for a panel primarily composed of _____ materials (usually wood) in the form of pieces or particles.

    a. cellulosic
    b. synthetic
    c. nylon
    d. sheathing

20. The following woods are defined as termite-resistant:

    a. redwood and eastern red cedar
    b. black locust and redwood
    c. cedar and redwood
    d. black walnut and red cedar

21. The grade of lumber is defined as:

    a. strength and durability
    b. strength and scent
    c. strength and utility
    d. utility and durability

!Codealert

A place of religious worship is a building or portion thereof intended for the performance of religious services.

!**Code**alert

A townhouse is a single-family dwelling unit constructed in a group of three or more attached units in which each unit extends from the foundation to the roof and has open space on at least two sides.

22. Black locust is which type of wood:

   a. nominal size   b. hardboard

   c. termite-resistant   d. decay-resistant

23. ANSI denotes which organization:

   a. American National Standards Installation

   b. American National Standards Institute

   c. American National Specifications Inc.

   d. American National Steel Institute

24. Which of the following best describes a townhouse?

   a. a single-family dwelling   b. a multifamily dwelling

   c. a castle   d. a landmark

25. Liquids that have a closed cup flash point at or above 200 degrees F are classified as:

   a. Class IIIB   b. Class IIB

   c. Class IB   c. Class IIA

TRUE-FALSE QUESTIONS

1. Aerosol products must be specified as Level 1, 2, or 3.

   True   False

2. Cryogenic fluid has a boiling point lower than 150 degrees F at 14.7 pounds per square inch.

   True   False

3. A helistop is the same as a heliport, except that no fuel, maintenance, repair, or storage is permitted.

True   False

4. HPM stands for Hazardous Production Material.

True   False

5. A basement must be considered as a story above the grade plane if the finished surface is less than 6 feet above grade plane.

True   False

6. An intermediate level or levels between the floor and ceiling of any story is the correct definition of mezzanine.

True   False

7. A fire partition is a horizontal assembly of materials designed to restrict the spread of fire by protecting openings.

True   False

8. The fire-separation distance is not to be measured from the building face to the side of the wall.

True   False

9. Construction documents can be written, graphic, or pictorial.

True   False

10. A vertical assembly is a fire-resistance-rated floor or roof assembly of materials designed to restrict the spread of fire.

True   False

11. A concrete-fiber blanket is a lightweight insulating material made of alumina-silica and shale.

True   False

12. Interior finish includes interior wall and roof finishes.

True   False

13. The definition of trim includes picture molds, chair rails, baseboards, and protective materials used in fixed application.

True   False

14.   A speaker or horn is part of an alarm-notification appliance.
True   False

15.   An egress court is a yard or court that provides access to a public way for one or more exits.
True   False

16.   Panic hardware that is listed for use on fire-door assemblies is called fire-entrance hardware.
True   False

17.   Nosing is defined as people who snoop around on building and construction sites.
True   False

18.   An entrance that is not a service entrance or a restricted entrance is called a public entrance.
True   False

19.   A scupper is a type of roof replacement.
True   False

20.   A diaphragm is the product of nominal strength and a resistance or strength-reduction factor.
True   False

21.   A factored load is defined as the product of a nominal load and a load factor.
True   False

22.   Load effects are conditions beyond which a structure becomes unfit for service and is judged to be no longer useful.
True   False

23.   To be in two horizontal directions at 90 degrees to each other is considered to be orthogonal.
True   False

24.   Sprayed fire-resistant materials are defined as fibrous materials that are applied by spraying to provide fire-resistant protection.
True   False

25.   Timber piles are made up of all the trees that have been cut from the construction or building site.
True   False

MULTIPLE-CHOICE ANSWER KEY

1. D	6. D	11. D	16. D	21. C
2. C	7. B	12. C	17. A	22. D
3. B	8. A	13. B	18. B	23. B
4. A	9. C	14. A	19. A	24. A
5. C	10. D	15. C	20. A	25. A

TRUE-FALSE ANSWER KEY

1. T	6. T	11. F	16. F	21. T
2. F	7. F	12. F	17. F	22. F
3. T	8. F	13. T	18. T	23. T
4. T	9. F	14. T	19. F	24. T
5. F	10. F	15. T	20. F	25. F

Chapter 3

USE AND OCCUPANCY CLASSIFICATIONS

As a contractor part of your responsibility is to know use and occupancy classifications of buildings and structures. If you are not aware of the purpose for which the room or space will be occupied, you could be in a world of trouble. Every group and/or occupancy type has its own code that must be applied. Certainly, codes written for Group A-1 occupancy are not going to work for Group R occupancies. This practice test is intended to help you discover how well you have read and understood Chapter 3, the use and occupancy classifications, of the 2006 International Building Code. So go ahead and take the test, and you will learn which areas need further study for the real test.

FILL-IN-THE-BLANK QUESTIONS

Choose the best word to complete the sentences below.

100	performing	congregate
International Fire Code	business	merchandise
50	S-2	transient
alcohol and drug	16	A-3
nightclubs	miscellaneous and utility	International Residential Code
F-2	moderate	combustible liquid
toxic	educational	dorms
boarding	U	jails
residential		

1.   _____ arts are classified as Group A-1 assemblies.

2.   Buildings such as courtrooms, lecture halls, and museums are classified as Group _____ assemblies.

3.   Group B is classified as: _____.

4.   High hazards such as _____ belong to Group H.

> ## !Codealert
> A building used for assembly purposes with an occupant load of less than 50 persons is classified as Group B occupancy.

> ## !Codealert
> A room or space that is used for assembly purposes with an occupant load of less than 50 persons and that is accessory to another occupancy shall be classified as a Group B occupancy or as part of that occupancy.

5. Group I 1 occupancy includes buildings or structures that house more than _____ people.

6. _____ and _____ centers are classified as Group I-1 occupancies.

7. Group M occupancies may include buildings for the display and sale of _____.

8. _____ is an occupancy of a dwelling or sleeping unit for not more than 30 days.

9. Leather is considered to belong to _____ hazard storage.

10. Group U is classified as _____ and _____.

11. Religious educational rooms that have an occupant load of fewer than _____ people must be classified as A-3 occupancies.

12. Factory industrial uses that are low-hazard must be classified as _____ occupancies.

> ## !Codealert
> A room or space used for assembly purposes that is less than 750 square feet in area and is accessory to another occupancy shall be classified as a Group B occupancy or as part of that occupancy.

!Codealert

High-hazard Group H classifications contain numerous exceptions and recent code changes.

13.   High-hazard Group H must be in accordance with the _____.

14.   Buildings that contain materials that are health hazards such as highly _____ materials must be classified as Group H-4.

15.   A building arranged or used for lodging for compensation, with or without meals, and not occupied as a single-family unit is a _____ house.

16.   A building that houses residents who do not require medical or nursing care is a_____ care service.

17.   _____ living facilities are building or parts of buildings that contain sleeping units for which residents share bathroom and/or kitchen facilities.

18.   Rooms in a building where students share sleeping facilities are called _____.

19.   Greenhouses are classified as Group _____.

20.   Adult and childcare facilities that are within a single-family home are to comply with the _____.

21.   A building used for assembly purposes with an occupancy load of fewer than _____ persons must be classified as Group B occupancy.

22.   Assembly uses intended for food and/or drink consumption include _____.

23.   Group I-3 occupancy includes buildings that are inhabited by more than five persons who are under constraint such as _____.

24. Group _____ includes buildings used for the storage of noncombustible materials such as glass or frozen foods.

25. Group E is classified as _____ occupancy.

TRUE-FALSE QUESTIONS

1. A separate single-story building without a basement or crawl space that is used for storage is a detached building.

True   False

2. Explosives can include dynamite igniters and fireworks.

True   False

3. Class V formulations burn intensely and pose a hazard.

True   False

4. An oxidizer is a material that gives off gas; chlorine is classified as an oxidizer.

True   False

5. A chemical that has a lethal dose of more than 50 milligrams per kilogram is considered toxic.

True   False

6. Pyrophoric is a chemical with an auto-ignition temperature in air at or below 110 degrees Fahrenheit.

True   False

!Codealert

Cotton made into banded bales with a packing density of at least 22 pounds per cubic foot and dimensions complying with the flooring: a length of 55 inches, a width of 21 inches, and a height of 27.6 to 35.4 inches.

> **!Code**alert
>
> Water-reactive material comes in different classes. Class 2 materials react violently with water or have the ability to boil water. Materials that produce flammable, toxic, or other hazardous gases or generate enough heat to cause autoignition or ignition of combustibles upon exposure to water or moisture.

7. Organic peroxides decompose into various unstable compounds over a long period of time.
    True   False

8. Materials that react explosively with water without heat or confinement are included in Class 3.
    True   False

9. A room or space that is used for assembly purposes, is less than 750 square feet in area, and is accessory to another occupancy must be classified as Group A.
    True   False

10. Amusement parks and post offices are classified in the same group.
    True   False

11. Corrosives include personal and household products.
    True   False

12. Stationary batteries used for facility emergency power are covered under the International Mechanical Code.
    True   False

> **!Code**alert
>
> R-3 use applies to adult facilities that provide accommodations for five or fewer persons of any age for less than 24 hours.

> **!Code**alert
>
> Congregate living facilities with 16 or fewer occupants are permitted to comply with the construction requirements for Group R-3.

13. An aerosol is a product that is dispensed from an aerosol container by a propellant.

    True   False

14. Aerosol products do not have a designated level.

    True   False

15. Baled cotton is a natural seed fiber wrapped in and secured with industry-accepted materials such as burlap.

    True   False

16. Densely packed baled cotton must be 55 inches in length, 21 inches wide, and 27.6 to 35.4 inches high.

    True   False

17. An artificial barricade is an artificial mound with a minimum thickness of 3 feet.

    True   False

18. Examples of closed systems for solids and liquids include product flow-through distillation.

    True   False

> **!Code**alert
>
> Congregate living facilities are buildings or parts thereof that contain sleeping units where residents share bathroom and or kitchen facilities.

!**Code**alert

"Transient" is described as occupancy of a dwelling or sleeping unit for not more than 30 days.

19. Day boxes are portable magazines designed to hold liquids such as water.
    True   False

20. Group U occupancies include greenhouses and carports.
    True   False

21. Low-hazard products are permitted to be wrapped in plastic.
    True   False

22. Buildings in Group I-3 are classified as one of five occupancy conditions.
    True   False

23. Group I, the institutional group, does not include assisted-living facilities.
    True   False

24. High-hazard Group H-1 is classified in divisions, such as explosives.
    True   False

25. Electrical coils are classified as storage, Group S-1.
    True   False

FILL-IN-THE-BLANK ANSWER KEY

1.	PERFORMING	13.	INTERNATIONAL FIRE CODE
2.	A-3	14.	TOXIC
3.	BUSINESS	15.	BOARDING
4.	COMBUSTIBLE LIQUIDS	16.	RESIDENTIAL
5.	16	17.	CONGREGATE
6.	ALCOHOL AND DRUG	18.	DORMS
7.	MERCHANDISE	19.	U
8.	TRANSIENT	20.	INTERNATIONAL RESIDENTIAL CODE
9.	MODERATE	21.	50
10.	MISCELLANEOUS AND UTILITY	22.	NIGHTCLUBS
		23.	JAILS
11.	100	24.	S-2
12.	F-2	25.	EDUCATIONAL

TRUE-FALSE ANSWER KEY

1.	T	6.	F	11.	F	16.	T	21.	T
2.	T	7.	T	12.	T	17.	T	22.	T
3.	F	8.	T	13.	T	18.	F	23.	F
4.	T	9.	F	14.	F	19.	F	24.	T
5.	T	10.	F	15.	T	20.	F	25.	F

Chapter 4

SPECIAL DETAILED REQUIREMENTS BASED ON USE AND OCCUPANCY

National Contractor's Exam Study Guide

This sample test regarding special detailed requirements based on use and occupancy is going to help you become better acquainted with the codes regarding special details that are sometimes overlooked in construction. Are you familiar with the occupant formula? Do you know what the smoke-control codes are? To which occupancy do they apply? These are just a few of the questions you will find on the sample test that you may be tested on later. There are a variety of questions ranging from exits to systems design. So long as you have studied, you should do well. On with the test!

MULTIPLE-CHOICE QUESTIONS

Choose the answer that best satisfies the questions below.

1. This is the occupant formula in for determining the required means of egress in a covered mall.
   a. $OMG = (0.0007)(GLA) + 25$
   b. $OLF = (0.00007)(GLA) + 25$
   c. $CME = (1.2345)(CME) + 30$
   d. $CME = (0.0007)(GLA) + 25$

2. The occupant load factor is not required to be less than _____ and must not be more than 50.
   a. 25   b. 30
   d. 15   d. 20

3. GLA stands for one of the following:
   a. good luck always   b. gross load area
   c. gross leasable area   d. gross less area

> **!Code**alert
>
> Where exit passageways provide a secondary means of egress from a tenant space, doorways to the exit passageway shall be protected by 1-hour fire-door assemblies that are self- or automatic-closing by smoke detection.

> **!Code**alert
>
> Children's playground structures have special requirements, and many of these requirements are new. See section 402.11 for complete details.

4. _____ and lobbies in Groups B, R-1, and R-2 are not required to comply with the provisions for covered mall buildings in Chapter 4.

   a. foyers

   b. decks

   c. ground-floor levels

   d. basements

5. Anchored buildings are not to be considered as part of :

   a. open parking garages

   b. covered parking garages

   c. playground structures

   d. covered mall buildings

6. OLF refers to:

   a. open load factor

   b. occupant load factor

   c. open lease factor

   c. occupant lease factor

7. Two exits must be provided for customers if the distance of travel is more than _____ feet.

   a. 50

   b. 55

   c. 75

   d. 65

> **!Code**alert
>
> Plastic signs affixed to the storefront of any tenant space facing a mall shall be limited as specified in Sections 402.15.1 through 402.15.5.2.

> **!Code**alert
>
> For buildings not greater than 420 feet in height, Type IA construction shall be allowed to be reduced to Type IB.

8.   All tenant areas, including areas used for storage, must be included in calculating the _____.

    a.   gross leasable area   b.   gross load area

    c.   gross load factor   d.   gross less area

9.   No modifications or changes in occupancy for a covered mall can be made without the approval of whom?

    a.   the mall owner   b.   the building official

    c.   the local town office   d.   the construction foreman

10.   A mall that ends at an anchor building where no other means of exit has been provided is considered?

    a.   a separate building   b.   a cul-de-sac

    c.   a barrier   d.   a dead-end

11.   Within each individual tenant space in a covered mall, the maximum distance of travel cannot be more than _____ feet.

    a. 150   b.   250

    c. 200   d.   225

> **!Code**alert
>
> In other than Group H occupancies, and where permitted by Exception 5 in Section 707.2, the provisions of this section shall apply to buildings or structures containing vertical openings defined herein as "atriums."

!**Code**alert

Smoke control is not required for atriums that connect only two stories.

12. 66 inches is the minimum width of:

a. a doorway

b. an exit passageway

c. a corridor

d. both b and c

13. Fire -_____ - rated separations are not required between tenant spaces and the mall.

a. resistance

b. reaction

c. proof

d. retardant

14. A small structure having one or more open sides, used as a newsstand or refreshment stand, is defined as a:

a. booth

b. stall

c. kiosk

d. stand

15. Sprinkler protection must be independent from systems provided for tenant spaces or _____.

a. garages

b. anchor stores

c. playgrounds

d. atriums

16. Children's playground structures may not be more than _____ square feet in area, unless a special investigation has shown satisfactory fire safety.

a. 300

b. 250

c. 325

d. 275

!**Code**alert

Vehicle ramps shall not be considered as required exits unless pedestrian facilities are provided.

> **!Code**alert
>
> Vehicle ramps that are utilized for vertical circulation as well as for parking shall not exceed a slope of 6.67 percent.

17. Horizontal sliding or _____ security doors must be in the full open position during business hours.

    a. fire-protection
    b. exit
    c. rolling
    d. vertical

18. Covered mall buildings more than 50,000 square feet in area must be provided with _____

    a. standby power
    b. smoke control
    c. both a and d
    d. an emergency voice/alarm communication system

19. _____ cannot take up more than 20 percent of the wall area facing the mall.

    a. doors
    b. signs
    c. banners
    d. windows

20. A building with an occupied floor located more than 75 feet above the lowest level is considered to be:

    a. a tower
    b. a high-rise
    c. a skyscraper
    d. a superstructure

> **!Code**alert
>
> Canopies used to support gaseous hydrogen systems require special consideration. See Section 406.5.2.1 for complete details.

> ## !Codealert
>
> Where a proscenium wall is required to have a fire-resistance rating, the stage opening shall be provided with a fire curtain of approved material or an approved water curtain complying with Section 903.3.1.1. The fire curtain shall be designed and installed to intercept hot gases, flames, and smoke and to prevent a glow from a severe fire on the stage from showing on the auditorium side for a period of 20 minutes. The closing of the fire curtain from the full open position shall be accomplished in less than 30 seconds, with the last 8 feet of travel requiring 5 or more seconds for full closure.

21. _____ control is not required for atriums that connect only two stories
    a. fire
    b. flame
    c. vapor
    d. smoke

22. The interior finish of walls and ceilings in an atrium cannot be less than Class _____ with no reduction in class for sprinkler systems.
    a. B
    b. C
    c. D
    d. II

23. Building spaces having a floor level used for human occupancy more than 30 feet below the lowest level are considered to be _____.
    a. secret spaces
    b. subterranean spaces
    c. underground buildings
    d. trains

> ## !Codealert
>
> Sprinklers are not required within portable orchestra enclosures on stages.

> **!Code**alert
>
> Aircraft hangars with individual lease spaces not exceeding 2,000 square feet each in which servicing, repairing, or washing is not conducted and fuel is not dispensed shall have floors that are graded toward the door but do not require a separator.

24.   Private garages and carports are classified as which group?

   a. Group A   b. Group B

   c. Group F   d. Group U

25.   Carports must have at least _____ open side(s).

   a. 2   b. 3

   c. 1   d. 4

TRUE-FALSE QUESTIONS

1.   Parking garages are classified as enclosed only.

   True   False

2.   Vehicle ramps are not required unless pedestrian facilities are provided.

   True   False

3.   Asphalt parking surfaces are permitted at ground level.

   True   False

4.   Parking garages are not to be considered separated from other occupancies.

   True   False

5.   Sleeping areas are permitted to have an opening from a parking garage.

   True   False

!**Code**alert

The floor construction of the control area and the construction supporting the floor of the control area is allowed to be 1-hour fire-resistance rated in buildings of Type IIA, IIIA, and VA construction, provided that both of the following conditions exist:

1. The building is equipped throughout with an automatic sprinkler system in accordance with Section 903.3.1.1.

2. The building is three stories, or less, in height.

6. Parking garages are not to be used as spaces to repair vehicles of any kind.

   True   False

7. A mechanical ventilation system must be provided for a parking garage.

   True   False

8. Repair garages must be constructed in accordance with the International Mechanical Code.

   True   False

9. Repair garages must be ventilated in accordance with the International Fire Code.

   True   False

!**Code**alert

When dealing with facilities for hazardous materials, walls shall not obstruct more than one side of a structure, except that walls shall be permitted to obstruct portions of multiple sides of the structure, provided that the obstructed area does not exceed 25 percent of the structure's perimeter.

> **!Code**alert
>
> Walls and floor/ceiling assemblies common to the room and to the building where the room is located shall be fire barriers, with not less than a 1-hour fire-resistance rating and without openings.

10. Institutional facility occupancies are classified as Group I-2.

    True   False

11. Spaces for doctors' and nurses' stations, charting, communications, and related clerical areas are permitted to be open to the corridor when constructed as such.

    True   False

12. Gift shops that are more than 500 square feet in area are permitted to be open to the corridor.

    True   False

13. Corridor walls are not required to be constructed as smoke partitions.

    True   False

14. Doors to resident sleeping units must have a clear width of not less than 28 inches.

    True   False

15. A fly gallery is a display of exotic flies found around the country.

    True   False

16. Combustible materials used in sets and sceneries on stages and platforms must meet the fire performance criteria of NFPA 701.

    True   False

17. Stages higher than 50 feet must be separated from dressing rooms, scene docks, and property rooms.

    True   False

18. Hazardous materials that are classified as Group H must comply with the International Hazard Code.

     True   False

19. Rooms, areas, or spaces in Group H in which explosives, corrosives, or combustibles are stored must be mechanically vented in accordance with both the International Fire Code and the International Hazard Code.

     True   False

20. Group H occupancies must be provided with an automatic fire-detection system.

     True   False

21. Flammable vapors or fumes are defined as a concentration of flammable materials exceeding 30 percent of their lower flammable unit.

     True   False

22. IDLH stands for "immediately dangerous to lives and health."

     True   False

23. Hazardous production material, or HPM, liquid is classified as either a Class I flammable liquid or a Class IIIA combustible liquid.

     True   False

24. Hydrogen cutoff rooms are not to be located below grade level.

     True   False

25. All hydrogen cutoff rooms must have mechanical ventilation and gas-detection systems connected to a standby power system.

     True   False

!Codealert

Section 420 pertaining to hydrogen cutoff rooms has numerous new regulations.

TABLE 4.1   Open parking garages area and height.

TYPE OF CONSTRUCTION	AREA PER TIER (square feet)	HEIGHT (in tiers)		
		Ramp access	Mechanical access Automatic sprinkler system	
			No	Yes
IA	Unlimited	Unlimited	Unlimited	Unlimited
IB	Unlimited	12 tiers	12 tiers	18 tiers
IIA	50,000	10 tiers	10 tiers	15 tiers
IIB	50,000	8 tiers	8 tiers	12 tiers
IV	50,000	4 tiers	4 tiers	4 tiers

TABLE 4.2   Explosion Control Requirements[a].

Material	Class	EXPLOSION CONTROL METHODS	
		Barricade construction	Explosion (deflagration) venting or explosion (deflagration) prevention systems[b]
HAZARD CATEGORY			
Combustible dusts[c]	—	Not Required	Required
Cryogenic flammables	—	Not Required	Required
Explosives	Division 1.1	Required	Not Required
	Division 1.2	Required	Not Required
	Division 1.3	Not Required	Required
	Division 1.4	Not Required	Required
	Division 1.5	Required	Not Required
	Division 1.6	Required	Not Required
Flammable gas	Gaseous	Not Required	Required
	Liquefied	Not Required	Required
Flammable liquid	IA[d]	Not Required	Required
	IB[e]	Not Required	Required
Organic peroxides	U	Required	Not Permitted
	I	Required	Not Permitted
Oxidizer liquids and solids	4	Required	Not Permitted
Pyrophoric gas	—	Not Required	Required
	4	Required	Not Permitted
Unstable (reactive)	3 Detonable	Required	Not Permitted
	3 Nondetonable	Not Required	Required
Water-reactive liquids and solids	3	Not Required	Required
	2[g]	Not Required	Required
SPECIAL USES			
Acetylene generator rooms	—	Not Required	Required
Grain processing	—	Not Required	Required
Liquefied petroleum gas-distribution facilities	—	Not Required	Required
Where explosion hazards exist[f]	Detonation	Required	Not Permitted
	Deflagration	Not Required	Required

a. See Section 414.1.3 of the *International Building Code 2006.*
b. See the *International Fire Code.*
c. As generated during manufacturing or processing,. See definition of "Combustible dust" in Chapter 3 of the *International Building Code 2006.*
d. Storage or use.
e. In open use or dispensing.
f. Rooms containing dispensing and use of hazardous materials when an explosive environment can occur because of the characteristics or nature of the hazardous materials or as a result of the dispensing or use process.
g. A method of explosion control shall be provided when Class 2 water-reactive materials can form potentially explosive mixtures.

MULTIPLE-CHOICE ANSWER KEY

1. B	6. B	11. C	16. A	21. D
2. B	7. C	12. D	17. D	22. A
3. C	8. A	13. A	18. C	23. C
4. A	9. B	14. C	19. B	24. D
5. D	10. D	15. B	20. B	25. A

TRUE-FALSE ANSWER KEY

1. F	6. T	11. T	16. T	21. T
2. T	7. T	12. F	17. T	22. F
3. T	8. F	13. F	18. F	23. T
4. F	9. F	14. T	19. F	24. T
5. F	10. T	15. F	20. F	25. T

Chapter 5

GENERAL BUILDING HEIGHTS AND AREAS

The height and area of different construction types must be governed by the intended use of the buildings and may not exceed the limits that the code has set forth. You, the contractor, have to be familiar with these code rules. It will not do you any good if you do not know the answers in this sample test regarding the codes for building heights and surrounding areas. Made up of fill-in-the-blank and true-false questions, you will be tested on your knowledge of unlimited-area buildings or perhaps mixed-use occupancy. Good luck!

FILL-IN-THE-BLANK QUESTIONS

Choose the best word to complete the sentences below.

1	Arabic	12
swimming	Type II	12
aircraft	7	mezzanine
60	elevated	stage
vertical	glazed	grade
alphabetical	S	separate
address	incinerator	

1.   Address numbers must be _____ numbers or _____ letters.

2.   The building height is the _____ distance from grade plane to the average height of the highest roof surface.

3.   An intermediate level or levels between the floor and ceiling of any story is defined as a _____.

4.   The area of a one-story, Group B, F, M, or _____ building is not to be limited when the building is equipped with an automatic sprinkler.

5.   The area of a one-story, group A-3 building is not to be limited if the building does not have a _____ other than a platform.

6.   A basement is considered to be a story above grade plane if the finished surface of the floor above it is more than _____ feet above the finished ground.

7.   _____ numbers must be a minimum of 4 inches high with a minimum width of 0.5 inch.

8.   Two or more buildings on the same lot are to be regulated as _____.

!**Code**alert

The aggregate area of mezzanines in buildings and structures of Type I or II construction shall not exceed one-half the area of the room in buildings and structures equipped throughout with an approved automatic sprinkler system in accordance with Section 903.3.1.1 and an approved emergency voice/alarm communication system in accordance with Section 907.2.12.2

9.   Motion-picture theaters are of a _____ construction.

10.   _____ rooms are required to have a 2-hour separation and an automatic sprinkler system.

11.   An _____ platform is an unoccupied platform used for mechanical systems.

12.   _____ plane is defined as a reference plane representing the average of finished ground levels adjoining the building at outside walls.

13.   The height of one-story _____ hangars is not limited if the building has an automatic fire-extinguishing system in accordance with the code.

14.   The clear height above and below the mezzanine-floor construction cannot be less than _____ feet.

15.   In industrial facilities, mezzanines used for control equipment are permitted to be _____ on all sides.

16.   The area of a one-story, Group F-2 or S-2 building is not limited when the building is surrounded and adjoined by public ways or yards no less than _____ feet wide.

17.   An automatic sprinkler system is not required for indoor sports, such as _____, provided that exit doors lead directly outside and there is a fire alarm system.

18.   Buildings must have approved address numbers in place so that they are _____.

> ## !Codealert
>
> In other than Groups H and I occupancies no more than two stories in height above grade plane and equipped throughout with an automatic sprinkler system in accordance with Section 903.3.1.1, a mezzanine having two or more means of egress shall not be required to be open to the room in which the mezzanine is located.

19. For a basement to be considered as a story, the finished surface of the floor above the basement must be more than _____ feet above the finished ground level.

20. Storage and laundry rooms both require a _____ -hour separation or an automatic fire-extinguishing system.

TRUE-FALSE QUESTIONS

1. Towers, spires, steeples, and other roof structures can be used as living space or storage.
   True   False

2. Roof structures are limited in height if they are made of combustible materials and are more than 20ft. tall.
   True   False

3. Equipment platforms and mezzanines include walkways, stairs, and ladders.
   True   False

4. Equipment platforms must be fully protected by sprinklers.
   True   False

5. Buildings do not have to connect to a public way to receive an increase in frontage.
   True   False

> ## !Codealert
> Mixed-use occupancy regulations in Section 508 have many new elements compared to previous code requirements.

6. Separate occupancies are individually classified.

   True   False

7. The height and number of floors above the basement in an enclosed parking garage are unlimited.

   True   False

8. Enclosed garages are allowed to have an office and/or waiting room.

   True   False

9. A building height is the horizontal distance from grade plane to the height of the highest roof surface.

   True   False

10. Special industrial occupancies are defined as buildings that require large areas and unusual heights for accommodation of machinery and equipment.

    True   False

> ## !Codealert
> Section 509.8 refers to Group B or M with Group S-2 open parking garage above has many new code changes.

TABLE 5.1   Incidental use areas.

ROOM OR AREA	SEPARATION AND/OR PROTECTION
Furnace room where any piece of equipment is over 400,000 Btu per hour input	1 hour or provide automatic fire-extinguishing system
Rooms with boilers where the largest piece of equipment is over 15 psi and 10 horsepower	1 hour or provide automatic fire-extinguishing system
Refrigerant machinery rooms	1 hour or provide automatic sprinkler system
Parking garage	2 hours or 1 hour and provide automatic fire-extinguishing system
Hydrogen cut-off rooms, not classified as Group H	1 hour in Group B, F, M, S and U occupancies 2 hour in Group A, E, I and R occupancies.
Incinerator rooms	2 hours and automatic sprinkler system
Pain shops, not classified as Group H, located in occupancies other than Group F	2 hours; or 1 hour and provide automatic fire-extinguishing system
Laboratories and vocational shops, not classified as Group H, located in Group E or I-2 occupancies	1 hour or provide automatic fire-extinguishing system
Laundry rooms over 100 square feet	1 hour or provide automatic fire-extinguishing system
Storage rooms over 100 square feet	1 hour or provide automatic fire-extinguishing system
Group I-3 cells equipped with padded surfaces	1 hour
Group I-2 cells waste and linen collection rooms	1 hour
Waste and linen collection rooms over 100 square feet	1 hour or provide automatic fire-extinguishing system
Stationary lead-acid battery systems having a liquid capacity of more than 100 gallons used for facility standby power, emergency power or uninterrupted power supplies	1 hour in Group B, F, M, S and U occupancies 2 hour in Group A, E, I and R occupancies

For SI: 1 square foot = 0.0929 m^2, 1 pound per square inch = 6.9 kPa, 1 British thermal unit per hour = 0.293 watts, 1 horsepower = 746 watts, 1 gallon = 3.785 L.

FILL-IN-THE-BLANK ANSWER KEY

1. ARABIC/ALPHABETICAL	8. SEPARATE	15. GLAZED
2. VERTICAL	9. TYPE II	16. 68
3. MEZZANINE	10. INCINERATOR	17. SWIMMING
4. S	11. ELEVATED	18. LEGIBLE
5. STAGE	12. GRADE	19. 12
6. 12	13. AIRCRAFT	20. 1
7. ADDRESS	14. 7	

TRUE-FALSE ANSWER KEY

1. F	3. F	5. F	7. T	9. F
2. F	4. T	6. T	8. T	10. T

Chapter 6
TYPES OF CONSTRUCTION

Welcome to the Chapter 6 sample test of the 2006 International Building Code. This test will check how well you know the provisions that control the classification of buildings by type of construction. Construction and building inspectors examine buildings, highways and streets, sewer and water systems, dams, bridges, and other structures to ensure that their construction, alteration, or repair complies with building codes and ordinances, zoning regulations, and contract specifications. Will your building or structure pass these tests? Come along and take the sample test on types of construction and the codes that pertain to them and find out how well you know the codes and which you need to understand better.

MULTIPLE-CHOICE QUESTIONS

Choose the answer that fits best in the statements below.

1. The 2006 Building Code contains _____ construction types.
    a. 3   b. 4
    c. 5   d. 2

2. Wood columns must be _____, laminated, or sawn and no less than 8 inches in any dimensions when used to support floor loads.
    a. glued   b. pasted
    c. attached by adhesive   d. compound

3. Wood beams or _____ must be sawn or glued-laminated timber and not be less than 6 inches in width or less than 10 inches in depth.
    a. poles   b. lathing
    c. girders   d. trusses

4. Framed timber trusses that support floor loads must not have members of less than _____ nominal inches in any dimension.
    a. 4   b. 8
    c. 10   d. 6

5. _____ plates cannot be less than 3 inches in thickness.
    a. splice   b. bond
    c. split   d. piece

> ## !Codealert
> In buildings of Type I construction exceeding two stories in height, fire-retardant-treated wood is not permitted in roof construction when the vertical distance from the upper floor to the roof is less than 20 feet.

6. Floors cannot have _____ spaces.

    a. hidden   b. open

    c. secret   d. concealed

7. _____ must be made of solid wood construction formed by not less than two layers of 1-inch boards.

    a. walls   b. floors

    c. partitions   d. roofs

8. The use of combustible piping materials is permitted when installed in accordance with which of the following groups of code:

    a. The International Mechanical and International Plumbing Codes

    b. The International Mechanical and International Fire Codes

    c. The International Plumbing and International Building Codes

    d. The International Fire and International Building Codes

9. When using electrical wiring methods, combustible insulation or tubing must be installed in accordance with which of the following:

    a. The International Fire Code

    b. The International Mechanical Code

    c. The International Electrical Wiring Code

    d. The ICC Electrical Code

10. Floors cannot extend closer than _____ to walls.

    a. 1 inch   b. 0.5 inch

    c. 2 inches   d. 1.5 inches

> **!Code**alert
>
> Sprayed fire-resistant materials and intumescent and mastic fire-resistant coatings, determined on the basis of fire-resistance tests in accordance with Section 703.2 and installed in accordance with Section 1704.10 and 1704.11, respectively.

TRUE-FALSE QUESTIONS

1. Fire-retardant-treated wood framing that complies with this code is permitted within exterior-wall assemblies of a 2-hour rating or less.

    True   False

2. Columns are to be continuous or superimposed and connected in an approved manner.

    True   False

3. It is OK to use Corbeling in place of molding for masonry walls under the floor.

    True   False

4. Columns and arches that conform to heavy timber sizes are not permitted to be used externally with a horizontal separation of 20 feet or more.

    True   False

5. Structural elements and exterior and interior walls in a Type II construction can be made of any materials.

    True   False

6. Fire-retardant-treated wood is permitted in nonbearing partitions where the required fire-resistance rating is 2 hours or less.

    True   False

7. An exception to question 6 exists in buildings of Type I, where fire-retardant-treated wood is not permitted if the vertical distance from the upper floor to the roof is less than 20 feet.

    True   False

8. Thermal and acoustical insulation, other than foam plastics, must have a flame-spread index of not more than 25.

    True   False

9. Roof coverings that have an A, B, or C classification are permitted to have thermal and acoustical insulation with a flame index of more than 25.

    True   False

10. In Type III construction the outside walls must be of noncombustible materials, and the inside building elements are of any materials.

    True   False

MULTIPLE-CHOICE ANSWER KEY

1. C	3. C	5. A	7. C	9. D
2. A	4. B	6. D	8. A	10. B

TRUE-FALSE ANSWER KEY

1. T	3. T	5. F	7. T	9. F
2. T	4. F	6. T	8. F	10. T

Chapter 7

FIRE-RESISTANCE-RATED CONSTRUCTION

Materials and assemblies used for structural fire resistance and fire-resistance-rated construction separation of adjacent spaces are some of the questions you will find in this sample test. How much of this will you know and recognize depends on how much effort you put into reading and getting to know the code book. This sample test has a little bit of everything. Your answers, whether right or wrong, can and should be used as a study guide to prepare yourself for the real test. Remember that in the real world, these tests are timed. While you don't want to whip through it without thoroughly reading each question, try not to linger too long on any one question. If time allows, you may always go back and answer the questions you have skipped.

FILL-IN-THE-BLANK QUESTIONS

Choose the best word to complete the sentences below.

100	performing	congregate
louvers	parapet	annular space
partition	wires	I-3
walls	International Mechanical Code	fire-protection-rating
25	T-rating	self
fire blocking	XXX	labels
horizontally	rolling	interstitial
joints	latch	type V
smoke detectors	membrane	side-hinged
ASTM E 119		

1.   _____ is the opening around the penetrating item.

2.   A _____ is the period of time that an opening protective assembly will maintain the ability to confine a fire.

3.   _____ is building material installed to resist the free passage of flames to areas of the building through concealed spaces.

!**Code**alert

A fire-resistance-rated wall assembly of materials designed to restrict the spread of fire in which continuity is maintained is a fire barrier.

> **!Code**alert
>
> Fire-separation distance is the distance measured from the building face to one of the following:
>
> - The closet interior lot line
> - To the centerline of a street, an alley, or public way
> - To an imaginary line between two buildings on the property

4. Shaft enclosures are the _____ or construction elements forming the boundaries of a shaft.

5. Openings in exterior walls must be separated _____ to protect against fire spread on the outside of the building.

6. A _____ is a type of barrier that must be provided on outside walls of buildings.

7. Except for _____ construction, fire walls must be made of noncombustible materials.

8. Fire-resistance-rated glazing must be tested in accordance with _____.

9. Walls separating dwelling units in the same building must comply with fire _____ codes.

> **!Code**alert
>
> A horizontal assembly is a fire-resistance-rated floor or roof assembly of materials designed to restrict the spread of fire in which continuity is maintained.

> **!Code**alert
>
> Mineral fiber is a type of insulation that is composed principally of fibers manufactured from rock, slag, or glass, with or without binders.

10. A 1-hour fire-resistance-rating is required for smoke barriers except for those barriers that are 0.10-inch-thick steel in Group _____ buildings.

11. Smoke-barrier walls are not required in _____ spaces that are designed and constructed with ceilings that provide resistance from fire smoke.

12. Doors in smoke partitions must not include _____.

13. Where required, doors in smoke partitions must be _____- or automatic-closing.

14. The space around penetrating items and in _____ must be filled with an approved material to limit the passage of smoke.

15. In 1-hour fire-resistance-rated floor construction, the ceiling _____ is not required to be installed over unusable crawl spaces.

> **!Code**alert
>
> Mineral wool is a synthetic vitreous fiber insulation made by melting predominantly igneous rock or furnace slag and other inorganic materials and then physically forming the melt into fibers.

> ## !Codealert
> Windows in exterior walls required to have protected openings in accordance with other sections of the code or determined to be protected in accordance with Section 704.3 or 704.8 shall comply with Section 715.5. Other openings required to be protected with fire-door or shutter assemblies in accordance with other sections of this code or determined to be protected in accordance with Section 704.3 or 704.8 shall comply with Section 715.4

16. Floor penetrations contained and located within the cavity of a wall do not require a _____.

17. Pipes, _____, and conduits must not be embedded in the required fire-protective covering of a structural member that is required to be encased.

18. _____ or pivoted swinging doors must be tested in accordance with NFPA 252 or UL 10C.

> ## !Codealert
> The fire-resistance rating of the fire barrier separating atriums shall comply with Section 404.5.

> ## !Codealert
> The fire barrier separating incidental-use areas shall have a fire-resistance rating of not less than that indicated in Table 508.2 of the code.

!Codealert

Fire barriers separating control areas shall have a fire-resistance rating of not less than that required in Section 414.2.3.

19. Fire doors must display _____ that must be permanently attached to the door or frame.

20. Unless otherwise permitted, single fire doors must have an active _____ bolt that secures the door when closed.

21. Doors through which pedestrians travel must be activated by _____ with alarm verification.

!Codealert

Where exterior walls serve as a part of a required fire-resistance-rated enclosure or separation, such walls shall comply with the requirements of Section 704 for exterior walls, and the fire-resistance enclosure or separation requirements shall not apply.

!Codealert

Penetrations in a fire barrier by ducts and air-transfer openings shall comply with Section 716.

!Codealert

In other than Group H occupancies, a shaft enclosure is not required for floor openings complying with the provision for atriums in Section 404.

22. _____ fire shutters must include automatic-closing devices.

23. The size limitation for a fire-protection-rating glazed window is _____ percent of the area of a common wall in any room.

24. _____ on a glazing label represents the fire-protection-rating period, in minutes, that was tested.

25. Fire dampers for hazardous exhaust systems must comply with the _____.

!Codealert

Code requirements for elevators have many updates, which you can find in Section 707.14 of the code.

!Codealert

Identification for fire-protection-rated glazing has current updates, which can be found in Section 715.6.3.1 of the code.

!Codealert

When dealing with the fire-resistance rating of structural members, the king studs and boundary elements that are integral elements in load-bearing walls of light-framed construction shall be permitted to have required fire-resistance ratings provided by the membrane protection provided for the load-bearing wall.

TRUE-FALSE QUESTIONS

1. Fire blocking is not required for slab-on-grade floors in gymnasiums.

   True   False

2. Draft stopping is required in buildings equipped with an automatic sprinkler system.

   True   False

3. Draft stopping doesn't have to be installed in concealed roof spaces as long as they are larger than 3,000 square feet.

   True   False

4. Unsanded gypsum is considered equal to gypsum sand plaster if mixed properly.

   True   False

!Codealert

Alternative methods for determining fire-protection ratings are new and can be found in Section 715.3 of the code.

!Codealert

See Section 715.4.3.2 of the code for updates on glazing in door assemblies.

5. Fiberboard insulation is the same as foam plastic insulation.

   True   False

6. A ceramic fiber blanket is defined as a lightweight insulating material consisting of inorganic glass fibers formed into rigid boards.

   True   False

7. The thickness of panels with tapered cross sections is determined at a distance of 6 inches from the edge.

   True   False

8. Joints between adjacent precast concrete slabs are to be considered in calculating the slab thickness provided that a concrete topping of at least 1 inch is used.

   True   False

9. F rating is the letter classification given to fire-resistance-rated materials.

   True   False

!Codealert

Smoke dampers are not required at penetrations of shafts where ducts are used as part of an approved mechanical smoke-control system designed in accordance with Section 909 and where the smoke damper will interfere with the operation of the smoke-control system.

> **!Code**alert
>
> Section 716.6.2 of the code deals with membrane penetrations and contains numerous changes in the code language.

10. A fire separation distance is the imaginary line to a street, alley, or public way.

   True   False

11. A window constructed and glazed to give protection against the passage of fire is a fire window assembly.

   True   False

12. An opening that passes through an entire assembly is called a passageway.

   True   False

13. Cornices, eave overhangs, and exterior balconies are just three examples of exterior walls.

   True   False

14. Type V construction is allowed for R-3 occupancies.

   True   False

15. Exterior walls must be of materials allowed by the building's type of construction.

   True   False

> **!Code**alert
>
> The fire resistance of structural steel trusses protected with intumescent or mastic fire-resistant coating shall be determined on the basis of fire-resistance tests in accordance with Section 703.2.

!**Code**alert

See Section 721.5.2.2.1 for changes in minimum thickness regulations.

16. A lightweight insulating concrete made by mixing preformed foam with Portland-cement slurry is considered to be perlite concrete.

    True   False

17. When rating structured-steel columns for fire resistance, the rating must be based on the size of the element and the type of protection.

    True   False

18. W/D means weight-to-heated perimeter ratios.

    True   False

19. When equating the fire resistance of structural steel, "R" refers to the fire-resistance rating of column assembly (in hours).

    True   False

20. It doesn't really matter how concrete is mixed or what it is mixed with as long as it suits your purposes.

    True   False

21. Fire-resistance ratings calculated using the code book must be used for 1-hour-rated assemblies.

    True   False

22. Membranes have assigned times on the fire-exposed sides.

    True   False

23. Allowable loads on beams and columns are determined using design values given in AF&PA NDS.

    True   False

24. Wood members do not have a minimum size requirement.

    True   False

25. The time assigned to double 3/8-inch gypsum wallboard is 40 minutes.

    True   False

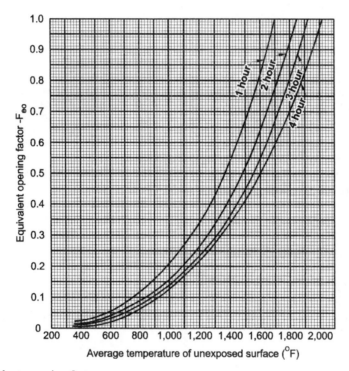

FIGURE 7.1   Equivalent opening factor.

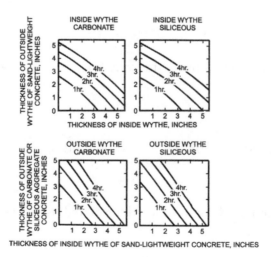

FIGURE 7.2   Fire-resistance ratings of two-wythe concrete walls.

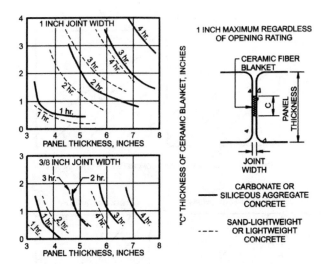

FIGURE 7.3 Ceramic fiber joint protection.

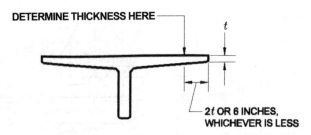

For SI: 1 inch = 25.4 mm.

FIGURE 7.4 Determination of slab thickness for sloping soffits.

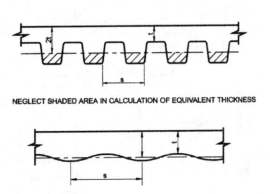

For SI: 1 inch = 25.4 mm.

FIGURE 7.5 Slabs with ribbed or undulating soffits.

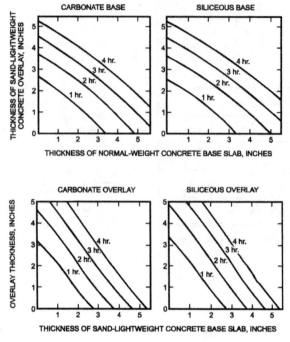

For SI: 1 inch = 25.4 mm.

FIGURE 7.6   Fire-resistance ratings for two-course concrete floors.

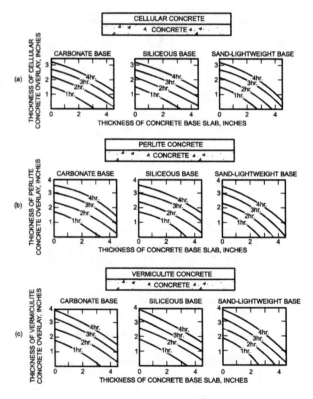

For SI: 1 inch = 25.4 mm.

FIGURE 7.7   Fire-resistance ratings for concrete roof assemblies.

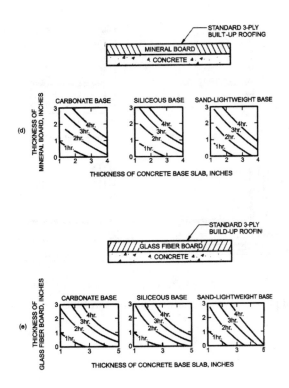

For SI: 1 inch = 25.4 mm.

FIGURE 7.8   Fire-resistance ratings for concrete.

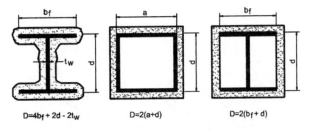

FIGURE 7.9   Determination of the heated perimeter of structural steel columns.

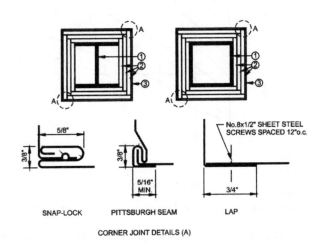

FIGURE 7.10   Gypsum wallboard protected structural steel columns with sheet steel column covers.

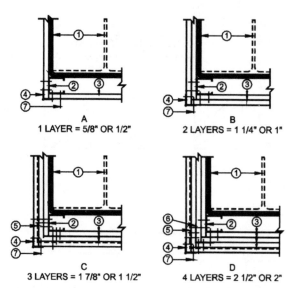

FIGURE 7.11   Gypsum wallboard protected structural steel columns with steel stud/screw attachment system.

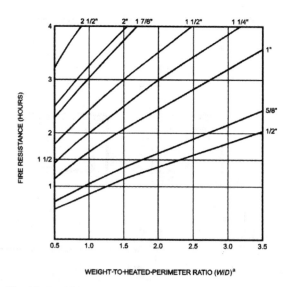

For SI: 1 inch = 25.4 mm, 1 pound per linear foot/inch = 0.059 kg/m/mm.

FIGURE 7.12   Fire resistance of structural steel columns protected with various thicknesses of Type X gypsum wallboard.

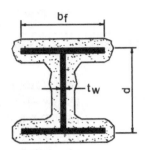

FIGURE 7.13   Wide flange structural steel columns with spray-applied fire-resistant materials.

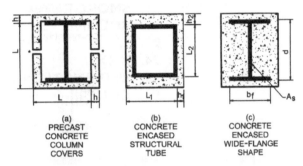

FIGURE 7.14   Concrete protected structural steel columns.

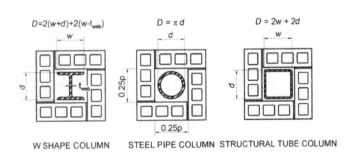

For SI: 1 inch = 25.4 mm.

FIGURE 7.15   Concrete or clay masonry protected structural steel columns.

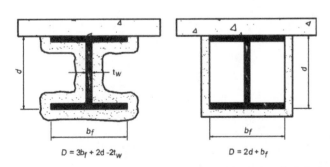

FIGURE 7.16   Determination of the heated perimeter of structural steel beams and girders.

FILL-IN-THE-BLANK ANSWER KEY

1. ANNULAR SPACE
2. FIRE PROTECTION RATING
3. FIRE BLOCKING
4. WALLS
5. HORIZONTALLY
6. PARAPET
7. TYPE V
8. ASTM E 119
9. PARTITION
10. I-3
11. INTERSTITIAL
12. LOUVERS
13. SELF
14. JOINTS
15. MEMBRANE
16. T-RATING
17. WIRES
18. SIDE-HINGED
19. LABELS
20. LATCH
21. SMOKE DETECTOR
22. ROLLING
23. 25
24. XXX
25. INTERNATIONAL MECHANICAL CODE

TRUE-FALSE ANSWER KEY

1. T	6. F	11. T	16. F	21. T
2. F	7. T	12. F	17. T	22. T
3. F	8. F	13. T	18. T	23. T
4. T	9. F	14. T	19. T	24. F
5. F	10. F	15. T	20. F	25. F

Chapter 8
INTERIOR FINISHES

In Chapter 8 of the 2006 International Building Code, you have learned about interior finishes, which include trim and wall or ceiling finishes. A building contractor has to know the code provisions while working with interior finishing. Without this knowledge how would you know what the flame-spread index is for walls and covered ceilings or what the ratings of combustible materials are? Although property owners may have an idea of what they want to use for interior trim, it is up to you to know if, according to the code, it may be used. If you haven't read and learned the code, you may install an item that, according to the building code, is not permitted. The Chapter 8 test includes multiple-choice and true-false questions about your working knowledge of interior finishes.

MULTIPLE-CHOICE QUESTIONS

1. Show windows in the exterior walls of the first story above grade shall _____ permitted to be of wood.

    a. be   b. not be

    c. occasionally be   d. none of the above

2. Show windows in the exterior walls of the first story above grade shall _____ permitted to be of unprotected metal framing.

    a. Be   b. not be

    c. Occasionally be   d. none of the above

3. The propagation of flame over a surface is known as _____.

    a. burnout   b. flame spread

    c. envelopment   d. none of the above

4. Interior finishes include _____.

    a. wall finishes   b. ceiling finishes

    c. floor finishes   d. all of the above*

5. The exposed floor surfaces of buildings, including coverings applied over a finished floor or stair, including risers, are known as _____.

    a. interior floor finishes   b. carpeting

    c. sub floors   d. none of the above

!Codealert

Show windows in the exterior walls of the first story above grade shall be permitted to be of wood or of unprotected metal framing.

6. Trim can include _____.
    a. picture molds
    b. headers
    c. chair rails
    d. either A or C

7. Trim can include _____.
    a. p molds
    b. baseboards
    c. chair rails
    d. all of the above

8. Trim can include _____.
    a. picture molds
    b. window frames
    c. chair rails
    d. all of the above*

9. Trim can include _____.
    a. door frames
    b. handrails
    c. chair rails
    d. all of the above*

!Codealert

Foam plastics shall not be used as interior finish or trim except as provided in Section 2603.9 or 2604. This section shall apply both to exposed foam plastics and to foam plastics used in conjunction with a textile or vinyl facing or cover.

!Codealert

A comparative measure, expressed as a dimensionless number, derived from visual measurements of the spread of flame versus time for a material tested in accordance with ASTM E 84.

10.   An interior wall or ceiling finish that is not more than _____ inch thick shall be applied directly against a noncombustible backing.

a.   0.25   b.   0.50

b.   0.65   d.   0.75

TRUE-FALSE QUESTIONS

1.   Combustible materials shall not be permitted for use as a finish for walls, ceilings, floors, and other interior surfaces of buildings.

True   False

2.   Foam plastics shall not be used as an interior finish or trim.

True   False

!Codealert

A comparative measure, expressed as a dimensionless number, derived from measurements of smoke obscuration versus time for a material tested in accordance with ASTM E 84.

3. Expanded-vinyl wall covering is a wall covering that consists of a woven textile backing, an expanded vinyl base-coat layer, and a nonexpanded vinyl skin coat.

    True   False

4. Combustible insulating boards are not permitted to be attached directly to a noncombustible floor assembly.

    True   False

5. Textile wall and ceiling coverings shall have a Class A flame-spread index.

    True   False

6. Interior-finish materials regulated by the code shall be applied or otherwise fastened in such a manner that such materials will not readily become detached where subjected to a room temperature of 180 degrees Fahrenheit.

    True   False

7. The stability of interior-finish materials must be capable of withstanding a temperature test for not less than 60 minutes.

    True   False

8. A Class C rating applies to materials with a flame-spread index of 76 to 200.

    True   False

9. A Class A rating applies to materials with a flame-spread index of 0 to 25.

    True   False

10. A Class B rating applies to materials with a flame-spread index of 26 to 75.

    True   False

!Codealert

See Section 803.6.2 for many changes in the code.

MULTIPLE-CHOICE ANSWER KEY

1. A	3. B	5. A	7. D	9. D
2. A	4. D	6. D	8. D	10. A

TRUE-FALSE ANSWER KEY

1. F	3. T	5. T	7. F	9. T
2. F	4. F	6. F	8. T	10. T

Chapter 9

FIRE-PROTECTION SYSTEMS

The key to passing the building-code exam is to learn it and to understand it. There is a big difference between learning the code and understanding it. Testing is designed to be tricky for the simple reason that it is designed to see if you have truly grasped the code. The sample test for Chapter 9, fire-protection systems, is one of these tests. With the abundance of definitions and the wealth of information in this chapter, there are undeniably a lot of code rules that must be understood. Understanding the code, which ranges from modifications to fire department hookups, is your responsibility. This sample test is designed to assist you in discovering which parts you need to study further. Good luck with this test, and I'll see you in Chapter 10.

MULTIPLE-CHOICE QUESTIONS

1. Automatic sprinkler systems shall be monitored by _____.
    a. the fire marshal
    b. a firefighter
    c. an approved supervising station
    d. none of the above

2. A fire-alarm-system component such as _____ or any combination thereof defines an alarm notification appliance.
    a. a bell   b. a horn
    c. both a and b   d. none of the above

3. A fire-alarm-system component such as _____ or any combination thereof defines an alarm notification appliance.
    a. a bell   b. a horn
    c. a light display   d. any of the above

4. A fire-alarm-system component such as _____ or any combination thereof defines an alarm notification appliance.
    a. a speaker   b. a text display
    c. both a and b   d. none of the above

5. An annunciator is a unit containing one or more indicator lamps, alphanumeric displays, or other equivalent means in which each indication provides status information about a _____.
    a. circuit   b. condition
    c. location   d. all of the above

6. Automatic, as applied to fire-protection devices, refers to a device or system providing an emergency function without the necessity for human intervention and activated as a result of a predetermined _____.

   a. time
   b. rate of temperature rise
   c. both a and b
   d. none of the above

7. Automatic, as applied to fire-protection devices, refers to a device or system that provides an emergency function without the necessity for human intervention and is activated as a result of a predetermined _____.

   a. temperature rise
   b. rate of temperature rise
   c. both a and b
   d. none of the above

8. Automatic sprinkler systems are usually activated by _____ and discharge water over the fire area.

   a. manual operation
   b. flame
   c. heat
   d. none of the above

9. Electrically nonconducting, volatile, or gaseous fire-extinguishing chemicals that do not leave a residue upon evaporation are known as _____.

   a. an automatic system
   b. a ceiling limit
   c. a clean agent
   d. none of the above

10. A fire command center is the principal _____ location where the status of detection, alarm communications, and control systems is displayed.

    a. attended
    b. unattended
    c. either a or b
    d. none of the above

11. A foam-extinguishing system is a special system discharging a foam made from concentrates spread _____ over the area to be protected.

    a. mechanically
    b. chemically
    c. either a or b
    d. none of the above

12. Halogenated extinguishing systems are fire-extinguishing systems using one or more atoms of an element from the halogen chemical series of _____.

    a. fluorine
    b. chlorine
    c. bromine
    d. all of the above

> **!Code**alert
>
> When approved by the fire-code official, mechanical smoke control for large enclosed volumes, such as atriums or malls, shall be permitted to utilize the exhaust method. Smoke-control systems using the exhaust method shall be designed in accordance with NFPA 92B.

13. Halogenated extinguishing systems are fire-extinguishing systems using one or more atoms of an element from the halogen chemical series of _____.

    a. iodine
    b. chlorine
    c. both a and b
    d. none of the above

14. A Class I standpipe system is one that provides _____ hose connections to supply water for use by fire departments and those trained in handling heavy fire streams.

    a. 1.5- inch
    b. 2-inch
    c. 2.5-inch
    d. 2.75-inch

15. A Class II standpipe system is one that provides _____ hose stations to supply water for use by the building occupants or by the fire department during initial response.

    a. 1.5- inch
    b. 2-inch
    c. 2.5-inch
    d. 2.75-inch

16. A Class III standpipe system is one that provides _____ hose connections to supply water for use by fire departments and those trained in handling heavy fire streams. In addition to the hose connections, a Class III system provides 1.5-inch hose stations for use by building occupants and the fire department.

    a. 1.5- inch
    b. 2-inch
    c. 2.5-inch
    d. 2.75-inch

17. Types of standpipe systems may include _____.

    a. automatic dry systems
    b. automatic grease systems
    c. automatic dense systems
    d. none of the above

18. Types of standpipe systems may include _____.
    a. automatic wet systems
    b. automatic grease systems
    c. automatic dense systems
    d. none of the above

19. Types of standpipe systems may include _____.
    a. manual dry systems
    b. manual wet systems
    c. either a or b
    d. none of the above

20. Types of standpipe systems may include _____.
    a. semiautomatic dry systems
    b. automatic grease systems
    c. automatic dense systems
    d. none of the above

21. The bulk storage of tires is described as tire storage where the area available for storage exceeds _____ cubic feet.
    a. 5,000
    b. 10,000
    c. 15,000
    d. 20,000

22. An automatic sprinkler system shall be provided throughout buildings and portions thereof used as _____ occupancies as provided for in the code.
    a. single-family
    b. dual-family
    c. group A
    d. none of the above

23. An automatic sprinkler system shall be provided for Group A-1 occupancies where the fire area exceeds _____ square feet.
    a. 8,000
    b. 9,500
    c. 10,000
    d. 12,000

24. An automatic sprinkler system shall be provided for Group A-1 occupancies where the fire area has an occupant load of _____ or more.
    a. 100
    b. 125
    c. 200
    d. 300

25. An automatic sprinkler system shall be provided for Group A-1 occupancies where the fire area is located on a _____other than the level of exit discharge.
    a. ceiling
    b. floor
    c. wall
    d. none of the above

26. An automatic sprinkler system shall be provided for Group A-1 occupancies where the fire area contains _____.

    a. a sleeping area
    b. a kitchen
    c. a multi-theater complex
    d. more than 10,000 square feet

27. An automatic sprinkler system shall be provided for Group A-2 occupancies where the fire area exceeds _____ square feet.

    a. 2,000
    b. 2,500
    c. 3,500
    d. 5,000

28. An automatic sprinkler system shall be provided for Group A-2 occupancies where the fire area has an occupant load of _____ or more.

    a. 100
    b. 125
    c. 200
    d. 300

29. An automatic sprinkler system shall be provided for Group A-5 occupancies in _____.

    a. concession stands
    b. retail areas
    c. press boxes
    d. all of the above

30. An automatic sprinkler system shall be provided for Group E occupancies where the fire areas are greater than _____ square feet.

    a. 10,000
    b. 15,000
    c. 20,000
    d. 25,000

31. An automatic sprinkler system shall be provided throughout all buildings containing a Group F-1 occupancy where the fire area exceeds _____ square feet.

    a. 9,000
    b. 12,000
    c. 14,000
    d. 18,000

!Codealert

The height of the lowest horizontal surface of the accumulating smoke layer shall be maintained at least 7 feet above any walking surface that forms a portion of a required egress system within the smoke zone.

32. An automatic sprinkler system shall be provided throughout all buildings containing a Group F-1 occupancy where the fire area is located more than _____ stories above grade.

    a. 2   b. 3

    c. 4   d. 6

33. An automatic sprinkler system shall be provided throughout all buildings containing a Group F-1 occupancy where fire areas on all floors, including mezzanines, exceed _____ square feet.

    a. 12,000   b. 15,000

    c. 20,000   d. 24,000

34. An automatic sprinkler system shall be provided throughout all buildings containing a Group F-1 occupancy where fire areas contain woodworking operations in excess of _____ square feet.

    a. 1,000   b. 1,500

    c. 2,000   d. 2,500

35. An automatic sprinkler system shall be provided in buildings, or portions thereof, where cellulose-nitrate film or pyroxylin plastics are manufactured, stored, or handled in quantities exceeding _____ pounds.

    a. 35   b. 75

    c. 100   d. 200

36. Where required by the International Mechanical Code, automatic sprinklers, with some exceptions, shall be provided in ducts conveying _____.

    a. warm air   b. cold air

    c. hazardous exhaust   d. all of the above

37. Where required by the International Mechanical Code, automatic sprinklers, with some exceptions, shall be provided in ducts conveying _____.

    a. flammable materials   b. combustible materials

    c. hazardous exhaust   d. all of the above

38. Not less than a _____-foot clearance shall be maintained between automatic sprinklers and the top of piles of combustible fibers.

    a. 1   b. 2

    c. 3   d. 4

39. Automatic sprinklers shall be installed in or under covered _____ that exceed 4 feet in width.

    a. kiosks   c. displays

    c. both a and b   d. none of the above

40. Automatic sprinklers shall be installed in or under covered _____ that exceed 4 feet in width.

    a. kiosks   b. booths

    c. both a and b   d. none of the above

41. Automatic sprinklers shall be installed in or under covered _____ that exceed 4 feet in width.

    a. concession stands   b. booths

    c. both a and b   d. none of the above

42. Automatic sprinklers shall be installed in or under covered _____ that exceed 4 feet in width.

    a. concession stands   b. equipment

    c. both a and b   d. none of the above

43. Automatic equipment interlocks for automatic fire systems may include _____.

    a. fuel shutoffs   b. door closers

    c. ventilation controls   d. all of the above

44. Automatic equipment interlocks for automatic fire systems may include _____.

    a. door openers   b. blind systems

    c. conveyor openings   d. all of the above

45. Automatic equipment interlocks for automatic fire systems may include _____.

    a. smoke vents   b. heat vents

    c. conveyor openings   d. all of the above

46. Construction documents for fire-alarm systems shall be submitted for review and approval prior to system installation. The documents shall include _____.

    a. annunciation   b. power connections

    c. battery calculations   d. all of the above

47. Where _____ conditions prohibit installation of automatic smoke detection, other automatic detection shall be allowed.

   a. weather

   b. wet

   c. ambient

   d. all of the above

48. A manual fire-alarm system shall be installed in Group B occupancies having an occupant load of _____ or more persons or more than 100 persons above or below the lowest level of exit discharge.

   a. 200

   b. 300

   c. 400

   d. 500

49. Buildings with a rating of Group I-3 shall be equipped with both manual and automatic fire-alarm systems for alerting _____.

   a. residents

   b. staff

   c. the fire department

   d. all of the above

50. Group R-1 occupancies require the installation of _____ fire-alarm system.

   a. a manual

   b. an automatic

   c. an audio

   d. none of the above

TRUE-FALSE QUESTIONS

1. Threads provided for fire-department connections to sprinkler systems, standpipes, yard hydrants, or any other fire-hose connection shall be compatible with the connections used by the local fire department.

   True   False

2. A supervising station is not required for automatic sprinkler systems protecting single-family homes.

   True   False

3. A supervising station is not required for automatic sprinkler systems protecting two-family homes.

   True   False

4.   Manual fire-alarm, automatic fire-extinguishing, and emergency-alarm systems in Group H occupancies shall be monitored by an approved supervising station.

      True   False

5.   An alarm signal is a signal indicating an emergency requiring immediate action.

      True   False

6.   Alarm verification features are designed to log in response time to emergencies.

      True   False

7.   A notification appliance that alerts through the sense of hearing is known as an audible alarm-notification appliance.

      True   False

8.   An automatic fire-extinguishing system is an approved system of devices and equipment that automatically detects a fire and sounds an alarm.

      True   False

9.   Carbon-dioxide extinguishing systems are systems that includes a manual or automatic actuating mechanism.

      True   False

10.   A deluge system is one in which, when a valve is opened, water flows into the piping system and discharges from all sprinklers attached thereto.

      True   False

11.   Heat is the energy produced by combustion that causes substances to drop in temperature.

      True   False

!Codealert

Fire-department connections are covered by Section 912. This section contains numerous code changes that need to be observed.

12. An effect produced by the sudden violent expansion of gases and accompanied by a shock wave, disruption of enclosing materials or structures, or both is known as a flash-out fire.

    True   False

13. A smoke detector is an example of an initiating device.

    True   False

14. A nuisance alarm can be an alarm that is activated by a cause that cannot be determined.

    True   False

15. A trouble signal is a signal initiated by the fire-alarm system or device indicative of a fire with flame spread.

    True   False

16. A wet-chemical extinguishing system is one that uses a solution of water and a potassium-carbonate-based chemical, a potassium-acetate-based chemical, or a combination thereof that forms an extinguishing agent.

    True   False

17. A zone is a defined area within the protected premises.

    True   False

18. An automatic sprinkler system shall be installed in Group H occupancies

    True   False

19. An automatic sprinkler system is not required in Group I-1 facilities.

    True   False

20. Areas used for the bulk storage of tires are required to be equipped with automatic sprinkler systems.

    True   False

21. Basements containing 1,500 square feet may be subject to the installation of an automatic sprinkler system.

    True   False

22. Kitchen exhaust hoods in commercial cooking operations are required to be equipped with an automatic sprinkler system.

    True   False

23. Sprinklers may be omitted from any room that is damp, is of fire-resistance-rated construction, or contains electrical equipment.

True   False

24. Fire-hose threads and fittings used in connection with automatic sprinkler systems shall be as prescribed by the local building official.

True   False

25. Approved audible deices shall be connected to every automatic sprinkler system.

True   False

26. Alarm devices shall be provided on the interior of buildings and are recommended on the exterior of buildings.

True   False

27. Approved control valves shall be provided at the point of connection to the riser on each floor in high-rise buildings.

True   False

28. Automatic sprinkler systems protecting commercial-type cooking equipment shall be supplied from a separate, readily accessible, indicating-type control valve that is identified.

True   False

29. Stages greater than 1,500 square feet in area shall be equipped with a Class III wet standpipe system.

True   False

30. Underground buildings shall be equipped throughout with a Class I automatic or manual wet standpipe system.

True   False

31. In Group I-1 buildings, corridors, habitable spaces other than sleeping units and kitchens, and waiting areas that are open to corridors shall be equipped with an automatic smoke-detection system.

True   False

32. Smoke detectors in high-rise buildings are to be connected to an automatic fire-alarm system.

True   False

33. Manual overrides for emergency voice communication shall not be allowed.

True   False

34. Covered mall buildings exceeding 35,000 square feet in total floor area shall be provided with an emergency voice-alarm communication system.

True   False

35. An approved automatic smoke-detection system shall be installed in areas containing stationary storage-battery systems having a liquid capacity of more than 50 gallons.

True   False

36. A presignal system shall not be installed unless approved by the fire-code official and the fire department.

True   False

37. Each floor shall be zoned separately, and a zone shall not exceed 20,000 square feet.

True   False

38. Visible alarm-notification appliances shall be provided in public areas and common areas.

True   False

39. Duct smoke detectors shall have an independent alarm that is not connected to the building's fire-alarm control panel.

True   False

40. Smoke-control systems are exempt from the building code.

True   False

41. Vestibule ventilation shall have a minimum net area of 8 square feet.

True   False

42. The vestibule ceiling shall be at least 24 inches higher than the door opening into a vestibule to serve as a smoke and heat trap.

True   False

43. Smoke and heat vents shall be capable of being operated by approved automatic and manual means.

True   False

44. The effective venting area for smoke and heat vents shall not be less than 16 square feet.

     True   False

45. A metal sign with raised letters at least 1 inch in size shall be mounted on all fire-department connections serving automatic sprinklers, standpipes, or fire pump connections.

     True   False

MULTIPLE-CHOICE ANSWER KEY

1.	C	11.	C	21.	D	31.	B	41.	C
2.	C	12.	D	22.	C	32.	B	42.	C
3.	D	13.	C	23.	D	33.	D	43.	D
4.	C	14.	C	24.	D	34.	D	44.	C
5.	D	15.	A	25.	B	35.	C	45.	D
6.	B	16.	C	26.	C	36.	C	46.	D
7.	C	17.	A	27.	D	37.	D	47.	C
8.	C	18.	A	28.	A	38.	C	48.	D
9.	C	19.	C	29.	D	39.	C	49.	B
10.	C	20.	A	30.	C	40.	C	50.	A

TRUE-FALSE ANSWER KEY

1.	T	10.	T	19.	F	28.	T	37.	F
2.	T	11.	F	20.	T	29.	F	38.	T
3.	T	12.	F	21.	T	30.	T	39.	F
4.	T	13.	T	22.	T	31.	T	40.	F
5.	T	14.	T	23.	F	32.	T	41.	F
6.	F	15.	F	24.	F	33.	F	42.	F
7.	T	16.	T	25.	T	34.	F	43.	T
8.	F	17.	T	26.	F	35.	T	44.	T
9.	T	18.	T	27.	T	36.	T	45.	T

Chapter 10

MEANS OF EGRESS

In Chapter 10, covering means of egress, there is a lot of information. All of it is extremely important for you to identify; otherwise how will you know which types of occupancies require certain types or a definite amount of egress? There are provisions in this chapter that control handrails—or maybe not? If you have studied this chapter, you already know the answer to this and many other questions in this sample test. Go ahead and take the test, and be sure to use your answers as a study guide for the real test. In the contracting world there is no such thing as knowing the code too well.

MULTIPLE-CHOICE QUESTIONS

1. An alternating tread device is one that has a series of steps between 50 and _____ degrees from horizontal, usually attached to a center support rail in an alternating manner so that the user does not have both feet on the same level at the same time.

    a. 60   b. 65

    c. 70   d. 75

2. Tiered seating facilities are known as _____.

    a. bleachers   b. mobile seating

    c. class II seating   d. none of the above

3. An enclosed exit-access component that defines and provides a path of egress travel to an exit is known as a _____.

    a. walkway   b. alley

    c. corridor   d. none of the above

!Codealert

A merchandise pad is an area for display of merchandise surrounded by aisles, permanent fixtures, or walls. Merchandise pads contain elements such as nonfixed and moveable fixtures, cases, racks, counters, and partitions from which customers browse or shop, as indicated in Section 105.2.

!Codealert

Check Section 1004 for code changes involving occupant load.

4.   A door equipped with double-pivoted hardware so designed as to cause a semi-counterbal-anced swing action when opening is known as _____.

a.   a swing door   b.   a balanced door

c.   an indirect door   d.   none of the above

5.   Panic hardware that is listed for use on fire-door assemblies is called _____.

a.   a push bar   b.   a push button

c.   fire-exit hardware   d.   none of the above

6.   The actual occupied area not including unoccupied accessory areas such as corridors, stairways, toilet rooms, mechanical rooms, and closets is known as the _____.

a.   gross floor area   b.   net floor area

c.   usable floor area   d.   none of the above

7.   Tiered seating facilities can be referred to as _____.

a.   folding seating   b.   telescopic seating

c.   a grandstand   d.   none of the above

!Codealert

Egress doors shall be side-hinged swinging, with some exceptions, one of which is a critical or intensive-care patient room within a suite of health-care facilities.

> **!Code**alert
>
> There are a number of code changes pertaining to exceptions for means of egress.

8. The leading edge of treads of stairs and of landings at the top of stairway flights is known as _____.

    a. edging   b. nosing

    c. quarter-round molding   d. none of the above

9. A ramp is a walking surface that has a running slope steeper than a _____-percent slope.

    a. 3   b. 5

    c. 7   d. 10

10. Two interlocking stairways providing two separate paths of egress located within one stairwell enclosure are known as _____.

    a. dual stairs   b. multi-stair assemblies

    c. scissor stairs   d. none of the above

11. The means of egress shall have a minimum ceiling height of _____.

    a. 7 feet   b. 7 feet 6 inches

    c. 8 feet   d. 10 feet

> **!Code**alert
>
> Stair riser heights shall be 7 inches maximum and 4 inches minimum. Additional code changes related to stair treads and risers can be found in Section 1009.3.

!**Code**alert

Outdoor stairways and outdoor approaches to stairways shall be designed so that water will not accumulate on walking surfaces.

12. Door closers and stops shall not reduce headroom to less than _____ inches.

    a. 60   b. 75

    c. 78   d. 80

13. A freestanding object mounted on a post or pylon shall not overhang that post or pylon more than _____ inches where the lowest point of the leading edge is more than 27 inches and less than 80 inches above the walking surface.

    a. 2   b. 4

    c. 6   d. 8

14. Where changes in elevation of less than _____ inches exist in the means of egress, sloped surfaces shall be used.

    a. 6   b. 10

    c. 12   d. none of the above

!**Code**alert

The walls and soffits within enclosed usable spaces under enclosed and unenclosed stairways shall be protected by 1-hour fire-resistance-rated construction or the fire-resistance rating of the stairway enclosure, whichever is greater. Access to the enclosed space shall not be directly from within the stair enclosure, with some exceptions.

!**Code**alert

Curved stairways with winder treads shall have treads and risers in accordance with Section 1009.3, and the smallest radius shall not be less than twice the required width of the stairway, with some exceptions.

15. When changes in elevation on a means of egress are _____ inches or less, the ramp shall be equipped with either handrails or floor-finish materials that contrast with adjacent floor-finish materials.

   a. 6   b. 10

   c. 12   d. none of the above

16. In order to be considered part of an accessible means of egress, an exit stairway shall have a clear width of _____ inches minimum between handrails and shall either incorporate an area of refuge within an enlarged floor-level landing or shall be accessed from either an area of refuge complying with Section 1007.6 or a horizontal exit.

   a. 24   b. 46

   c. 42   d. 48

17. Areas of refuge shall be provided with a(n) _____ system between the area of refuge and a center control point.

   a. alarm   b. phone

   c. two-way-communication   d. none of the above

!**Code**alert

Stairways shall have handrails on each side and shall comply with Section 1012. Where glass is used to provide the handrail, the handrail shall also comply with Section 2407 and pertinent exceptions.

> **!Code**alert
> Where the roof-hatch opening providing the required access is located within 10 feet of the roof edge, such roof access or roof edge shall be protected by guards installed in accordance with the provisions of Section 101.3.

18. The exterior area for assisted rescue shall be at least _____ percent open, and the open area above the guards shall be so distributed as to minimize the accumulation of smoke or toxic gases.

    a. 25   b. 45

    c. 50   d. 65

19. The minimum width of each door opening shall be sufficient for the occupant load thereof and shall provide a clear width of not less than _____ inches.

    a. 30   b. 32

    c. 36   d. 42

20. Projections into clear width areas shall not be lower than _____ inches above the floor or ground.

    a. 26   b. 30

    c. 32   d. 36

> **!Code**alert
> The floor or ground surface of a ramp run or landing shall extend 12 inches minimum beyond the inside face of a handrail complying with Section 1012.

!Codealert

The word "EXIT" shall be in high contrast with the background and shall be clearly discernible whether the means of exit-sign illumination is or is not energized. If a chevron directional indicator is provided as part of the exit sign, the construction shall be such that the direction of the chevron directional indicator cannot be readily changed.

21. Revolving doors shall not be given credit for more than _____ percent of the required egress capacity.

    a. 20   b. 25

    c. 40   d. 50

22. Revolving doors shall be credited with no more than a ____-person occupancy.

    a. 20   b. 25

    c. 40   d. 50

23. Each revolving door shall be capable of being collapsed when a force of not more than _____ pounds is applied within 3 inches of the outer edge of a wing.

    a. 75   b. 100

    c. 130   d. 150

24. Thresholds at doorways shall not exceed _____ inch in height for sliding doors serving dwelling units.

    a. 0.25   b. 0.5

    c. 0.75   d. 1

25. Space between two doors in a series shall be _____ inches minimum, plus the width of a door swinging into the space.

    a. 24   b. 30

    c. 36   d. 48

26. Horizontal sliding or swinging gates exceeding the 4-foot maximum leaf-width limitation are permitted in _____.

    a. fences
        b. walls surrounding a stadium

    c. both a and b
        d. none of the above

27. The width of a stairwell shall not be less than _____ inches.

    a. 30
        b. 36

    c. 42
        d. 44

28. Stair risers shall have a maximum height of _____ inches.

    a. 5
        b. 6

    c. 7
        d. 8

29. Stair risers shall have a minimum height of _____ inches.

    a. 4
        b. 6

    c. 7
        d. 8

30. When exceptions are applied, spiral stairways complying with Section 1009.8 are permitted a _____-inch headroom clearance.

    a. 72
        b. 74

    c. 78
        d. 80

!Codealert

Section 1014, dealing with exit access, has many recent code updates that must be considered.

TRUE-FALSE QUESTIONS

1.   Accessible means of egress is a continuous and unobstructed way of travel from any accessible point in a building or facility to a public way.

     True   False

2.   An exit-access component that defines and provides a path of egress travel is an aisle.

     True   False

3.   An area where persons unable to use stairways can remain temporarily to await instructions or assistance during emergency evacuation is a refuge area.

     True   False

4.   A court or yard that provides access to a public way with no more than one exit is known as an egress court.

     True   False

5.   An operable window shall not be considered to be an emergency escape and rescue opening.

     True   False

6.   An exit discharge is that portion of a means-of-egress system between the termination of an exit and a public way.

     True   False

7.   An exit access is that portion of a means-of-egress system that leads from any occupied portion of a building or structure to an exit.

     True   False

> ## !Codealert
> Where access to three or more exits is required, at least two exit doors or exit-access doorways shall be arranged in accordance with the provisions of Section 1015.2.1.

!**Code**alert

See Section 1020, on vertical exit enclosures, for code changes and additions.

8.   A handrail is a vertical rail intended for grasping by the hand for guidance or support.

True   False

9.   A merchandise pad is an area for display of merchandise surrounded by aisles, permanent fixtures, or walls.

True   False

10.   Smoke-protected assembly seating is seating served by a means of egress that is not subject to smoke accumulation within or under a structure.

True   False

11.   A change in elevation consisting of one or more risers is known as a stair.

True   False

12.   One or more flights of stairs, either exterior or interior, with the necessary landings and platforms connecting them to form a continuous and uninterrupted passage from one level to another is a stairway.

True   False

13.   A winder is a tread with non-parallel edges.

True   False

14.   A stairway having an open circular form in its plan view with uniform section-shaped treads is known as a spiral stairway.

True   False

15.   A public way is required to have a clear width and height of not less than 12 feet.

True   False

16.   Protruding objects shall not reduce the minimum clear width of accessible routes.

       True   False

17.   Walking surfaces of the means of egress shall have a slip-resistant surface and be securely attached.

       True   False

18.   Escalators, since they can be walked upon without electrical power, shall be considered a means of egress.

       True   False

19.   The posting of allowable occupant load is required in a place of assembly occupancy.

       True   False

20.   Each door providing access to an area of refuge from an adjacent floor area shall be identified by a sign.

       True   False

21.   Egress doors shall not be side-hinged swinging doors.

       True   False

22.   There shall be a floor or landing on each side of a door.

       True   False

23.   Bolt locks are prohibited.

       True   False

24.   The unlatching of any door or leaf shall not require more than two operations.

       True   False

!Codealert

With some exceptions, the common path of egress travel shall not exceed 30 feet from any seat to a point where an occupant has a choice of two paths of egress travel to two exits.

25. Locks may be permitted to prevent operation of doors at places of detention or restraint.

True   False

26. Turnstiles that restrict travel to one direction shall not be placed so as to obstruct any required means of egress.

True   False

27. Winder treads are prohibited in all stairways used as a means of egress.

True   False

28. Outdoor stairways must be constructed to prevent the accumulation of water on walking surfaces.

True   False

29. Ramps with a rise greater than 6 inches shall have handrails on both sides.

True   False

30. The minimum headroom in all parts of a means-of-egress ramp shall not be less than 78 inches.

True   False

TABLE 10.1   Maximum floor area allowances per occupant.

FUNCTION OF SPACE	FLOOR AREA IN SQ. FT. PER OCCUPANT
Accessory storage areas, mechanical equipment room	300 gross
Agricultural building	300 gross
Aircraft hangers	500 gross

Airport terminal
    Baggage claim
    Baggage handling
    Concourse
    Waiting area
	
	20 gross
	300 gross
	100 gross
	15 gross

Assembly Gaming floors (keno, slots, etc.)	 11 gross
Assembly with fixed seats	See Section 1004.7 *International Building Code 2006*

Assembly without fixed seats
    Concentrated (chairs only—not fixed)
    Standing space
    Unconcentrated (tables and chairs)
	
	7 net
	5 net
	15 net

Bowling centers, allow 5 persons for each lane including 15 feet of runway, and for additional areas	 7 net
Business areas	100 gross
Courtrooms—other than fixed seating areas	40 net
Day care	35 net
Dormitories	50 gross

Educational
    Classroom area
    Shops and other vocational room areas
	
	20 net
	50 net

Exercise rooms	50 gross
H-5 Fabrication and manufacturing areas	200 gross
Industrial areas	100 gross

Institutional areas
    Inpatient treatment areas
    Outpatient areas
    Sleeping areas
	
	240 gross
	100 gross
	120 gross

Kitchens, commercial	200 gross

Library
    Reading rooms
    Stack area
	
	50 net
	100 gross

Locker rooms	50 gross

Mercantile
    Areas on other floors
    Basement and grade floor areas
    Storage, stock, shipping areas
	
	60 gross
	30 gross
	300 gross

Parking garages	200 gross
Residential	200 gross

Skating rinks, swimming pools
    Rink and pool
    Decks
	
	50 gross
	15 gross

Stages and platforms	15 net
Warehouses	500 gross

For SI: 1 square foot = 0.0929 m²

MULTIPLE-CHOICE ANSWER KEY

1.	C	7.	C	13.	B	19.	B	25.	D
2.	A	8.	B	14.	C	20.	C	26.	C
3.	C	9.	B	15.	A	21.	D	27.	D
4.	B	10.	C	16.	D	22.	D	28.	C
5.	C	11.	B	17.	C	23.	C	29.	A
6.	B	12.	C	18.	C	24.	C	30.	C

TRUE-FALSE ANSWER KEY

1.	T	7.	T	13.	T	19.	T	25.	T
2.	T	8.	F	14.	F	20.	T	26.	T
3.	T	9.	T	15.	F	21.	F	27.	F
4.	F	10.	T	16.	T	22.	T	28.	T
5.	F	11.	T	17.	T	23.	T	29.	T
6.	T	12.	T	18.	F	24.	F	30.	F

Chapter 11
ACCESSIBILITY

The design and construction of facilities for accessibility is the subject of Chapter 11. A relatively short chapter but packed with must-know information; it pertains to the groups and occupancies that must abide by the accessibility codes. If you have studied this chapter well, you should have no problem passing this sample test regarding accessibility.

MULTIPLE-CHOICE QUESTIONS

1. A site, building, facility, or portion thereof that complies with the accessibility chapter of the code is considered to be _____.

   a. accessible
   b. readily accessible
   c. both a and b
   d. none of the above

2. A continuous, unobstructed path that complies with the accessibility chapter of the code is considered to be _____.

   a. accessible
   b. readily accessible
   c. an accessible route
   d. none of the above

3. An exterior or interior way of passage for pedestrians from one place to another is called a _____.

   a. bridge
   b. tunnel
   c. crosswalk
   d. circulation path

4. Interior or exterior rooms, spaces, circulation paths, or elements that are not for public use and are made available for the shared use of two or more people are known as _____.

   a. accessible
   b. facility
   c. common use
   d. none of the above

!Codealert

A dwelling unit or sleeping unit must comply with this code and the provisions for accessible units in ICC A 117.1.

> ## !Codealert
> All or any portion of a space used only by employees and only for work. Corridors, toilet rooms, kitchenettes, and break rooms are not employee work areas.

5.   Areas that do not qualify as employee work areas include _____.

   a. corridors   b.   toilet rooms

   c. both a and b   d.   none of the above

6.   Areas that do not qualify as employee work areas include _____.

   a. corridors   b.   toilet rooms

   c. break rooms   d.   all of the above

7.   Areas that do not qualify as employee work areas include _____.

   a. kitchenettes   b.   toilet rooms

   c. break rooms   d.   all of the above

8.   All or any portion of buildings, structures, site improvements, elements, and pedestrian or vehicular routes located on a site can be called _____.

   a. a facility   b.   an area

   c. an open space   d.   none of the above

> ## !Codealert
> A dwelling unit or sleeping unit with habitable space located on more than one story is a multistory unit.

!Codealert

A parcel of land bounded by a lot line or a designated portion of a public right-of-way is a site.

9.   A restricted entrance is one that is made available for common use on a controlled basis but not for _____.

a. private use   b. public use

c. licensed use   d. none of the above

10.   A restricted entrance is one that is made available for common use on a controlled basis but not for a _____.

a. private entrance   b. public entrance

c. service entrance   d. none of the above

11.   A service entry is an entrance intended primarily for delivery of _____.

a. goods   b. services

c. either a or b   d. none of the above

12.   Detached _____ dwellings and accessory structures and their associated sites are not required to be accessible.

a. one-family   b. two-family

c. both a and b   d. none of the above

!Codealert

A Type A unit is a dwelling unit or sleeping unit designed and constructed for accessibility in accordance with this code and the provisions for Type A units in ICC A117.1.

!**Code**alert

A Type B unit is a dwelling unit or sleeping unit designed and constructed for accessibility in accordance with this code and the provision for Type B units in ICC A117.1, consistent with the design and construction requirements of the federal Fair Housing Act.

13. Construction sites, structures, and equipment directly associated with the actual process of construction including but not limited to _____ are not required to be accessible.

    a. scaffolding

    b. bridging

    c. either a or b

    d. none of the above

14. Construction sites, structures, and equipment directly associated with the actual process of construction including but not limited to _____ are not required to be accessible.

    a. materials hoists

    b. bridging

    c. either a or b

    d. none of the above

15. Construction sites, structures, and equipment directly associated with the actual process of construction including but not limited to _____ are not required to be accessible.

    a. materials hoists

    b. materials storage

    c. construction trailers

    d. all of the above

16. Raised areas used primarily for purposes of security, life safety, or fire safety including but not limited to _____ are not required to be accessible or to be served by an accessible route.

    a. security

    b. life safety

    c. observation galleries

    d. all of the above

17. Raised areas used primarily for purposes of security, life safety, or fire safety including but not limited to _____ are not required to be accessible or to be served by an accessible route.

    a. fire towers

    b. life safety

    c. prison guard towers

    d. all of the above

> **!Code**alert
>
> Where required, sites, buildings, structures, facilities, elements, and spaces, temporary or permanent, shall be accessible to persons with physical disabilities.

18. Raised areas used primarily for purposes of security, life safety, or fire safety including but not limited to _____ are not required to be accessible or to be served by an accessible route.

    a. fire towers   b. lifeguard stands

    c. prison guard towers   d. all of the above

19. Nonoccupiable spaces accessed only by _____ are not required to be accessible.

    a. crawl spaces   b. catwalks

    c. either a or b   d. none of the above

20. Nonoccupiable spaces accessed only by _____ are not required to be accessible.

    a. freight elevators   b. catwalks

    c. either a or b   d. none of the above

21. Nonoccupiable spaces accessed only by _____ are not required to be accessible.

    a. very narrow passageways   b. ladders

    c. either a or b   d. none of the above

> **!Code**alert
>
> Walk-in coolers and freezers intended for employee use only are not required to be accessible.

22.   Spaces frequented only by personnel for maintenance, repair, or monitoring of equipment are not required to be accessible. Such spaces may include _____.

   a.   elevator pits

   b.   elevator penthouses

   c.   either a or b

   d.   none of the above

23.   Spaces frequented only by personnel for maintenance, repair, or monitoring of equipment are not required to be accessible. Such spaces may include _____.

   a.   mechanical rooms

   b.   elevator penthouses

   c.   electrical rooms

   d.   all of the above

24.   Spaces frequented only by personnel for maintenance, repair, or monitoring of equipment are not required to be accessible. Such spaces may include _____.

   a.   communications equipment rooms

   b.   meeting rooms

   c.   churches

   d.   all of the above

> **!Code**alert
>
> Accessible units and Type B units shall be provided in general-purpose hospitals, psychiatric facilities, detoxification facilities, and residential-care/assisted-living facilities of Group I-2 occupancies in accordance with Sections 1107.5.3.1 and 1107.5.3.2.

25. Spaces frequented only by personnel for maintenance, repair, or monitoring of equipment are not required to be accessible. Such spaces may include _____.

    a. restaurants

    b. meeting rooms

    c. churches

    d. none of the above

26. Spaces frequented only by personnel for maintenance, repair, or monitoring of equipment are not required to be accessible. Such spaces may include _____.

    a. piping catwalks

    b. meeting rooms

    c. churches

    d. none of the above

27. Spaces frequented only by personnel for maintenance, repair, or monitoring of equipment are not required to be accessible. Such spaces may include _____.

    a. piping catwalks

    b. equipment catwalks

    c. either a or b

    d. none of the above

28. Spaces frequented only by personnel for maintenance, repair, or monitoring of equipment are not required to be accessible. Such spaces may include _____.

    a. water-treatment rooms

    b. sewage-treatment rooms

    c. either a or b

    d. none of the above

29. Spaces frequented only by personnel for maintenance, repair, or monitoring of equipment are not required to be accessible. Such spaces may include _____.

    a. water-treatment pump rooms

    b. sewage-treatment pump rooms

    c. either a or b

    d. none of the above

> # !Codealert
> Accessible units shall be provided in Group I-3 occupancies in accordance with Sections 1107.5.5.1 through 1107.5.5.3.

30. Spaces frequented only by personnel for maintenance, repair, or monitoring of equipment are not required to be accessible. Such spaces may include _____.

    a. electric substations

    b. transformer vaults

    c. either a or b

    d. none of the above

31. Spaces frequented only by personnel for maintenance, repair, or monitoring of equipment are not required to be accessible. Such spaces may include _____.

    a. tunnel utility facilities

    b. highway utility facilities

    c. either a or b

    d. none of the above

32. In multilevel buildings and facilities, at least _____ accessible route(s) shall connect each accessible level, including mezzanines.

    a. 1

    b. 2

    c. 3

    d. 4

33. Where provided, _____ access for pedestrians from parking structures to building or facility entrances shall be accessible.

    a. direct

    b. indirect

    c. open

    d. restricted

> # !Codealert
> Accessible units and Type B units shall be provided in Group R-4 occupancies in accordance with Sections 1107.6.4.1 and 1107.6.4.2.

> **!Code**alert
>
> With one exception, at least one wheelchair space shall be provided in team or player seating areas serving areas of sport activity.

34. With some exceptions, at least one accessible entrance shall be provided to each _____, dwelling unit, and sleeping unit in a facility.

    a. door   b. office

    c. tenant   d. none of the above

35. Hospital parking areas are required to offer _____ percent of patient and visitor parking spaces provided to serve hospital outpatient facilities with accessibility.

    a. 5   b. 10

    c. 15   d. 25

36. Rehabilitation facilities and outpatient physical-therapy facilities shall have a minimum of one accessible parking space with _____ percent of the total parking area being accessible.

    a. 5   b. 10

    c. 15   d. 20

37. For every _____ or fraction of _____ accessible parking spaces, at least one shall be a van-accessible parking space.

    a. two   b. four

    c. six   d. eight

38. Continuous loading zones are required where passenger loading zones are provided; one passenger loading zone in every continuous _____ linear feet maximum of loading-zone space shall be accessible.

    a. 50   b. 75

    c. 100   d. 150

39. A passenger loading zone shall be provided at an accessible entrance to licensed medical and long-term-care facilities where people receive physical or medical treatment or care and where the period of stay exceeds _____ hours.

    a. 8   b. 12
    c. 16   d. 24

40. Nursing homes require at least _____ percent but not less than one of each type of dwelling and sleeping units to be accessible.

    a. 20   b. 25
    c. 40   d. 50

41. At least one wheelchair space shall be provided in _____ seating areas, with the exception of seating areas serving bowling lanes or other areas of sport activity.

    a. team   b. player
    c. either a or b   d. none of the above

42. When working with designated aisle seats, at least _____ percent but not less than one of the total number of aisle seats provided shall be designated aisle seats.

    a. 5   b. 10
    c. 12   d. 15

43. Accessible self-service storage facilities that have between 1 and 200 storage units are required to have _____ percent of the spaces but not less than 1 to be accessible.

    a. 5   b. 10
    c. 12   d. 15

> ## !Codealert
> Lawn-seating areas and exterior-overflow seating areas, where fixed seats are not provided, shall connect to an accessible route.

44. Where water-closet compartments are provided in a toilet room or bathing facility, the code requires a minimum of a _____.

    a. wheelchair-accessible compartment

    b. smoke detector

    c. both a and b

    d. none of the above

45. When drinking fountains are provided on an exterior site, on a floor, or within a secured area, no fewer than _____ drinking fountains shall be provided.

    a. two   b. three

    c. four   d. five

46. Self-service shelves and display units shall be located on _____.

    a. each floor of a multilevel building

    b. the floor

    c. an accessible route

    d. both a and c

47. Passenger transit platform edges bordering a dropoff and not protected by platform screens or guards shall have _____.

    a. rails   b. lights

    c. a detectable warning   d. all of the above

48. Where dressing rooms, fitting rooms, or locker rooms are provided, at least _____ percent but not less than one of each type of use in each cluster provided shall be accessible.

    a. 5   b. 10

    c. 15   d. 25

49. Food-service lines shall be accessible. Where self-service shelves are provided, at least _____ percent but not less than one of each type provided shall be accessible.

    a. 20   b. 25

    c. 40   d. 50

50. Raised boxing or wrestling rings _____ required to be accessible.

    a. are   b. are not

    c. must be   d. none of the above

TRUE-FALSE QUESTIONS

1. An accessible unit is a dwelling or sleeping unit that complies with the code and the provisions for accessibility.

    True   False

2. A detectable warning is a device used to warn visually impaired persons of hazards on a circulation path.

    True   False

3. Multilevel assembly seating is arranged in distinct levels where each level is comprised of either multiple rows or a single row of box seats accessed from a separate level.

    True   False

4. A dwelling or sleeping unit with habitable space located on more than one story is known as a multistory unit.

    True   False

5. An entrance that functions as a service entrance is known as a public entrance.

    True   False

6. A site is a parcel of land bounded by a lot line or a designated portion of a public right-of-way.

    True   False

7. A space for a single wheelchair and its occupant is known as a wheelchair space.

    True   False

!Codealert

Where drinking fountains are provided on an exterior site, on a floor or within a secured area, the drinking fountains shall be provided in accordance with Sections 1109.5.1 and 1109.5.2.

8. Spaces frequented only by personnel for maintenance, repair, or monitoring of equipment are required to be accessible.

    True   False

9. Toll booths are required to be accessible.

    True   False

10. If a daycare facility is part of a dwelling unit, only the portion of the structure utilized for the daycare facility is required to be accessible.

    True   False

11. In detention and correctional facilities, common-use areas that are used only by inmates or detainees and security personnel and that do not serve accessible holding or housing cells are not required to be accessible.

    True   False

12. In detention and correctional facilities, common-use areas that are used only by inmates or detainees and security personnel, and that do not serve accessible holding cells or housing cells are required to be served by an accessible route.

    True   False

13. Walk-in coolers and freezers intended for employee use only are not required to be accessible.

    True   False

14. At least one accessible route shall connect accessible buildings, accessible facilities, accessible elements, and accessible spaces that are on the same site.

    True   False

15. An accessible route is not required between accessible buildings, accessible facilities, accessible elements, and accessible spaces that have, as the only means of access between them, a vehicular way that does not provide for pedestrian access.

    True   False

16. With some exceptions, common-use circulations paths within employee work areas shall be accessible routes.

    True   False

17. Where direct access is provided for pedestrians from a pedestrian tunnel or elevated walkway to a building or facility, at least two entrances to the building or facility from each tunnel or walkway shall be accessible.

    True   False

18. Where restricted entrances are provided to a building or facility, at least one such entrance to the building or facility shall be accessible.

True   False

19. Where entrances used only by inmates or detainees and security personnel are provided at judicial, detention, or correctional facilities, at least one such entrance shall be accessible.

True   False

20. Service entrances shall be exempt from accessibly under all conditions.

True   False

21. An accessible entrance is not required for tenants in buildings not required to be accessible.

True   False

22. Accessible parking spaces shall be located on the shortest accessible route of travel from adjacent parking to an accessible building entrance.

True   False

23. A passenger loading zone shall be provided at locations of valet parking services.

True   False

24. At least one accessible route shall connect accessible building or facility entrances with the primary entrance of each accessible unit.

True   False

25. Lawn seating areas and exterior overflow seating areas, where fixed seats are not provided, shall not connect to an accessible route.

True   False

!**Code**alert

See Section 1109.13 and Section 1109.14 for recent changes in the code.

MULTIPLE-CHOICE ANSWER KEY

1. A	11. C	21. C	31. C	41. C
2. C	12. C	22. C	32. A	42. A
3. D	13. C	23. D	33. A	43. A
4. C	14. C	24. A	34. C	44. A
5. C	15. D	25. D	35. B	45. A
6. D	16. D	26. A	36. D	46. C
7. D	17. D	27. C	37. C	47. C
8. A	18. D	28. C	38. C	48. A
9. B	19. C	29. C	39. D	49. D
10. C	20. C	30. C	40. D	50. B

TRUE-FALSE ANSWER KEY

1. T	6. T	11. T	16. T	21. T
2. T	7. T	12. F	17. F	22. T
3. T	8. F	13. T	18. T	23. T
4. T	9. F	14. T	19. T	24. T
5. F	10. T	15. T	20. F	25. F

Chapter 12

INTERIOR ENVIRONMENT

National Contractor's Exam Study Guide

Interior environment can include anything from ventilation to interior spaces of buildings. If you put time and effort into studying Chapter 12, you should obtain a passing grade on this sample test. Do you know the codes as well as you think you should? Are there certain subjects in regard to interior environment that you want to know better? By taking this sample test, you can see how much you really know and what you need to read again. This test is designed to give you a taste of the real thing and to give you an edge. So what are you waiting for—go for it!

TRUE-FALSE QUESTIONS

1. A one-story addition to an existing building with a glazing area in excess of 30 percent of the gross area of the structure's exterior walls and roof is a sunroom addition.

   True   False

2. A one-story addition to an existing building with a glazing area in excess of 40 percent of the gross area of the structure's exterior walls and roof is a sunroom addition.

   True   False

3. A two-story addition to an existing building with a glazing area in excess of 40 percent of the gross area of the structure's exterior walls and roof is a sunroom addition.

   True   False

4. A separation of conditioned spaces between a sunroom addition and a dwelling unit consisting of existing or new wall(s), doors, and/or windows is known as thermal isolation.

   True   False

5. Enclosed attics and enclosed rafter spaces formed where ceilings are applied directly to the underside of roof framing members shall have cross-ventilation for each separate space by means of ventilating openings protected against the entrance of rain and snow.

   True   False

6. Blocking and bridging in an enclosed attic shall be arranged so as not to interfere with the movement of air.

   True   False

7. A minimum of 2 inches is required between attic insulation and roof sheathing.

   True   False

8. The net free ventilating area of an enclosed attic shall not be less than 1/150 of the area of the space ventilated, with 50 percent of the required ventilating area provided by ventilators located in the upper portion of the space to be ventilated at least 3 feet above eave or cornice vents and the balance of the required ventilation provided by eave or cornice vents.

    True   False

9. Exterior openings into attic space of any building intended for human occupancy shall be covered with corrosion-resistant wire cloth screening, hardware cloth, perforated vinyl, or similar material.

    True   False

10. The space between the bottom of the floor joists and the earth under any building, except space occupied by a basement or cellar, shall be provided with ventilation openings through foundation or exterior walls.

    True   False

11. Ventilation in under floor applications is prohibited from causing cross ventilation.

    True   False

12. The minimum net area of ventilation openings for under floor ventilation shall not be less than 1 square foot for each 200 square feet of crawl-space area.

    True   False

13. Natural ventilation of an occupied pace shall be through windows, doors, louvers, or other openings to the outdoors.

    True   False

14. Operating mechanisms for natural-ventilation openings shall be provided with ready access.

    True   False

15. The minimum openable area to the outdoors for natural ventilation shall be 8 percent of the floor area being ventilated.

    True   False

16. Where rooms and spaces without openings to the outdoors are ventilated through an adjoining room, the opening to the adjoining room shall be unobstructed and shall have an area of not less than 4 percent of the floor area of the interior room or space and not less than 25 square feet.

    True   False

17. Where openings below grade provide required natural ventilation, the outside horizontal clear space measured perpendicular to the opening shall be one and one-half times the depth of the opening.

    True   False

18. Rooms containing bathtubs, showers, spas, and similar bathing fixtures shall be mechanically ventilated in accordance with the International Mechanical Code and the International Fire Code.

    True   False

19. Interior spaces intended for human occupancy shall be provided with active or passive space-heating systems.

    True   False

20. Ratings for temperature control in buildings intended for human occupancy require heating systems to maintain a minimum indoor temperature of 72 degrees F at a point 3 feet above the floor on the design heating day.

    True   False

21. When used to meet requirements for natural light, the minimum net glazed area shall not be less than 8 percent of the floor area of the room served.

    True   False

22. For the purpose of natural lighting, any room may be considered as a portion of an adjoining room where one-half of the area of the common wall is open and unobstructed and provides an opening of not less than one-tenth of the floor area of the interior room or 25 square feet, whichever is greater.

    True   False

23. Artificial light shall be provided that is adequate to provide an average illumination of 8 foot-candles over the area of the room at a height of 30 inches above the floor level.

    True   False

24. Stairways within dwelling units and exterior stairways serving a dwelling unit shall have an illumination level on the tread run of not less than 1 footcandle.

    True   False

25. Yards for one- and two-story buildings must have a minimum width of 5 feet.

    True   False

26. For buildings more than two stories in height, the minimum width of the yard shall be increased at the rate of 1 foot for each additional story up to 14 stories.

True   False

27. When dealing with court areas, access shall be provided to the bottom of courts for cleaning purposes.

True   False

28. Courts more than two stories in height shall be provided with a vertical air intake at the bottom not less than 10 square feet in area.

True   False

29. The bottom of every court shall be properly graded and drained.

True   False

30. Habitable spaces other than kitchens shall not be less than 8 feet in any plan dimension.

True   False

31. Kitchens shall have a clear passageway of not less than 3 feet between counter fronts and appliances or counter fronts and walls.

True   False

32. Occupiable spaces, habitable spaces, and corridors shall have a ceiling height of not less than 8 feet.

True   False

33. Bathrooms may have a minimum ceiling height of 7 feet.

True   False

34. Kitchens may have a minimum ceiling height of 7 feet.

True   False

35. Laundry rooms may have a minimum ceiling height of 6 1/2 feet.

True   False

36. Storage rooms may have a minimum ceiling height of 6 1/2 feet.

True   False

37. Any room with a furred ceiling shall be required to have the minimum ceiling height in one-third of the area served.

True   False

38. Every dwelling unit shall have at least one room that shall have not less than 120 square feet of net floor area.

True   False

39. Every kitchen in one- and two-family dwellings shall have not less than 50 square feet of gross floor area.

True   False

40. Crawl spaces shall be provided with a minimum of one access opening not less than 18 inches by 24 inches.

True   False

41. An opening not less than 18 inches by 24 inches shall be provided to any attic area.

True   False

42. A 30-inch minimum of clear headroom shall be provided at or above the access opening in an attic space.

True   False

43. In units other than dwelling units, toilet and bathing-room floors shall have a smooth, hard, nonabsorbent surface that extends upward onto the walls at least 6 inches.

True   False

44. Walls within 3 feet of urinals and water closets shall have a smooth, hard, nonabsorbent surface, to a height of 4 feet above the floor, and except for structural elements the material used in such walls shall be of a type that is not adversely affected by moisture.

True   False

45. Shower compartments and walls above bathtubs with installed shower heads shall be finished with a smooth, nonabsorbent surface to a height not less than 72 inches above the drain inlet.

True   False

46. Built-in tubs with showers shall have waterproof joints between the tub and the adjacent wall.

True   False

47. Any room used for the preparation of food for service to the public shall have a toilet room that opens directly into the food-preparation area.

    True   False

48. Grab bars in bathrooms are required to be installed and sealed to protect structural elements from moisture.

    True   False

49. An efficiency dwelling unit shall have a living room of not less than 220 square feet of floor area.

    True   False

50. Efficiency dwelling units are required to have a minimum of one separate closet.

    True   False

TRUE-FALSE ANSWER KEY

1. F	11. F	21. T	31. T	41. F
2. T	12. F	22. T	32. F	42. T
3. F	13. T	23. F	33. T	43. T
4. T	14. T	24. T	34. T	44. F
5. T	15. F	25. F	35. F	45. F
6. T	16. F	26. T	36. F	46. T
7. F	17. T	27. T	37. F	47. F
8. T	18. T	28. F	38. T	48. T
9. T	19. T	29. T	39. T	49. T
10. T	20. F	30. F	40. T	50. T

Chapter 13
ENERGY EFFICIENCY

Chapter 13, which covers energy efficiency, was brief and did not warrant a question and answer section.

Chapter 14
EXTERIOR WALLS

There are many amendments made to the International Building Code year after year; it is highly recommended that you stay on top of these changes. A missed change in the code can mean loss of time and money and a stop order on your current project. In Chapter 14 you will find such changes for exterior walls in 2006. Stay on top of these code changes and you will do well on this test. Go ahead—see for yourself.

MULTIPLE-CHOICE QUESTIONS

1. Veneer secured and supported through the adhesion of an approved bonding material applied to an approved backing creates an _____ masonry veneer.

   a. altered

   b. adhered

   c. alternative

   d. none of the above

2. Veneer secured with approved mechanical fasteners to an approved backing is known as _____ masonry veneer.

   a. anchored

   b. adhered

   c. both a and b

   d. none of the above

3. The wall or surface to which veneer is secured is known as _____.

   a. the main wall

   b. the key wall

   c. backing

   d. none of the above

4. An exterior wall is a wall, bearing or nonbearing, that is used as an enclosure for a building, other than a fire wall, and that has a slope of _____ degrees or greater with the horizontal plane.

   a. 45

   b. 50

   c. 60

   d. 75

5. Metal composite material systems consist of an exterior-wall finish system fabricated using MCM in a specific assembly including _____.

   a. joints

   b. seams

   c. both a and b

   d. none of the above

6. Metal composite material systems consist of an exterior-wall finish system fabricated using MCM in a specific assembly including _____.

    a. attachments   b. seams

    c. both a and b   d. none of the above

7. Metal composite material systems consist of an exterior-wall finish system fabricated using MCM in a specific assembly including _____.

    a. attachments   b. substrate

    c. both a and b   d. none of the above

8. Metal composite material systems consist of an exterior-wall finish system fabricated using MCM in a specific assembly including _____.

    a. metal skins   b. seams

    c. both a and b   d. none of the above

9. Metal composite material systems consist of an exterior-wall finish system fabricated using MCM in a specific assembly including _____.

    a. framing   b. seams

    c. both a and b   d. none of the above

10. Veneer is a facing attached to a wall for the purpose of providing _____.

    a. ornamentation   b. strength

    c. both a and b   d. none of the above

11. Veneer is a facing attached to a wall for the purpose of providing _____.

    a. protection or insulation   b. strength

    c. both a and b   d. none of the above

!Codealert

Vinyl siding is a shaped material, made principally from rigid polyvinyl chloride (PVC), that is used as an exterior wall covering.

> ## !Codealert
> A material behind an exterior wall covering that is intended to resist liquid water that has penetrated behind the exterior covering from further intrusion into the exterior-wall assembly is a water-resistive barrier.

12. Exterior-wall-envelope test assemblies shall be at least 4 feet by _____ feet in size.

    a. 2   b. 4

    c. 6   d. 8

13. Exterior-wall-envelope assemblies shall be tested at a minimum differential pressure of _____ pounds per square foot.

    a. 5   b. 6

    c. 6.24   d. 6.5

14. Exterior-wall-envelope assemblies shall be subjected to a minimum test exposure duration of _____ hours.

    a. 1 1/4   b. 1 1/2

    c. 2   d. 2 1/2

15. Masonry walls require _____ to be installed in the first course above the finished ground level above the foundation wall or slab.

    a. flashing   b. weep holes

    c. both a and b   d. none of the above

16. Stone veneer units not exceeding _____ inches in thickness shall be anchored directly to masonry, concrete, or stud construction.

    a. 6   b. 10

    c. 12   d. 16

17. Veneer units of marble, travertine, granite, or other stone units of slab form ties of corrosion-resistant dowels in drilled holes shall be located in the middle third of the edge of the units and paced a maximum of _____ inches apart around the periphery of each unit with not less than four ties per veneer unit.

    a. 12

    b. 16

    c. 18

    d. 24

18. Interior adhered masonry veneers shall have a maximum weight of _____ psf.

    a. 10

    b. 15

    c. 20

    d. 25

19. Metal veneers that are not constructed of approved corrosion-resistant materials shall be protected front and back with _____.

    a. paint

    b. grease

    c. porcelain enamel

    d. none of the above

20. Exterior metal veneer shall be securely attached to the supporting masonry or framing members with _____.

    a. corrosion-resistant fastenings

    b. metal ties

    c. approved devices

    d. any of the above

21. Methods for protecting metal supports for exterior metal veneer against weather include _____.

    a. painting

    b. galvanizing

    c. either a or b

    d. none of the above

22. The areas of a single section of thin exterior structural-glass veneer shall not exceed _____ square feet where the section is not more than 15 feet above the level of a sidewalk or grade level directly below.

    a. 6

    b. 8

    c. 10

    d. none of the above

23. The areas of a single section of thin exterior structural-glass veneer shall not exceed _____ square feet where the section is more than 15 feet above the level of a sidewalk or grade level directly below.

    a. 6   b. 8

    c. 10   d. either a or b

24. The length or height of any section of thin exterior structural-glass veneer shall not exceed _____ inches.

    a. 16   b. 24

    c. 36   d. 48

25. Where thin exterior structural-glass veneer is installed above the level of the top of a bulk-head facing or at a level more than _____ inches above the sidewalk level, the mastic cement binding shall be supplemented with approved nonferrous metal shelf angles located in the horizontal joints in every course.

    a. 12   b. 24

    c. 32   d. 36

26. Special tests or calculations may be required for vinyl siding installed in regions where wind speeds exceed _____ miles per hour.

    a. 50   b. 75

    c. 100   d. none of the above

27. Special tests or calculations may be required for vinyl siding installed on buildings where the building height exceeds _____ feet.

    a. 24   b. 32

    c. 40   d. none of the above

28. Nails used to fasten vinyl siding shall be corrosion-resistant and shall be long enough to penetrate the studs or nailing strip by at least _____ inch.

    a. 1/4   b. 1/2

    c. 3/4   d. 1

29. When installing fiber cement siding on metal framing, all-weather screws shall be used and must penetrate the metal framing at least _____ full threads.

    a. two   b. three

    c. four   d. six

!Codealert

See Section 1405.12.2 for code updates on window sills.

30. Weather boarding and wall covers shall be securely fastened with _____ or other approved corrosion-resistant fasteners.

    a. copper   b. zinc

    c. either a or b   d. none of the above

31. Weather boarding and wall covers shall be securely fastened with _____ or other approved corrosion-resistant fasteners.

    a. zinc-coated   b. zinc

    c. either a or b   d. none of the above

32. Weather boarding and wall covers shall be securely fastened with _____ or other approved corrosion-resistant fasteners.

    a. aluminum   b. zinc

    c. either a or b   d. none of the above

33. Fiber-cement siding panels shall be installed with the _____ dimension parallel to the framing.

    a. nominal   b. short

    c. long   d. none of the above

34. Horizontal joints for fiber-cement siding shall be _____.

    a. flashed with Z-flashing

    b. covered

    c. blocked with solid wood framing

    d. both a and c

35. Horizontal lap siding for fiber-cement siding shall be lapped a minimum of _____ inches.

    a. 1 1/4   b. 1 1/2

    c. 1 3/4   d. 2

36.   Lap siding for fiber-cement siding shall be permitted to be installed with the fastener heads
      _____.

      a.   dimpled   b.   concealed

      c.   exposed   d.   both b and c

37.   Where installed on exterior walls having a fire-separation distance of ____ feet or less, com-
      bustible exterior wall coverings shall not exhibit sustained flaming.

      a.   3   b.   5

      c.   8   d.   10

38.   The acronym MCM represents _____.

      a.   metal composite materials

      b.   molded composite materials

      c.   metal component materials

      d.   none of the above

39.   MCM shall not be installed more than ____ feet in height above the grade plane.

      a.   16   b.   24

      c.   32   d.   40

40.   Where the fire-separation distance is 5 feet or less, the area of MCM shall not exceed _____
      percent of the exterior-wall surface.

      a.   10   b.   15

      c.   20   d.   25

!Codealert
The plastic core of the MCM shall not contain foam-plastic insulation as defined in Section 2602.1.

TRUE-FALSE QUESTIONS

1. An exterior-wall envelope is a system or assembly of exterior-wall components, including finish materials, that provides protection of the building structural members, including framing and sheathing materials and conditioned interior space, from the detrimental effects of the exterior environment.

    True   False

2. Fiber cement siding is a manufactured, fiber-reinforcing product made with an inorganic hydraulic or calcium silicate binder formed by chemical reaction and reinforced with organic or inorganic non-asbestos fibers or both.

    True   False

3. Metal composite material is a factory-manufactured panel consisting of metal skin bonded to one face of a plastic core.

    True   False

4. Veneer is a facing material used to increase the strength of a wall.

    True   False

5. A shaped material, made principally from rigid polyvinyl chloride, that is used as an exterior-wall covering is known as vinyl siding.

    True   False

6. Water-resistive barriers are installed in front of an exterior wall.

    True   False

7. Exterior walls shall provide buildings with weather-resistant envelopes.

    True   False

8. An exterior-wall envelope shall not include flashing.

    True   False

9. Flashing shall be installed in such a manner as to prevent moisture from entering a wall or to redirect it to the exterior.

    True   False

10. Flashing shall not be installed under sills.

    True   False

11. Copings require flashing that has projecting flanges.
     True   False

12. Exterior-wall pockets must be protected from accumulating water.
     True   False

13. Weep holes are required in masonry exterior walls.
     True   False

14. Anchor ties for stone veneer must be corrosion-resistant wire or an approved equivalent.
     True   False

15. The legs of loops in anchor ties shall be not less than 8 inches in length.
     True   False

16. Slab-type veneer units not exceeding 2 inches in thickness shall be anchored directly to masonry, concrete, or stud construction.
     True   False

17. Terra cotta or ceramic units not less than 1 5/8 inches thick shall be anchored directly to masonry, concrete, or stud construction.
     True   False

18. Metal veneers must be, without exception, fabricated from approved corrosion-resistant materials.
     True   False

19. Masonry backup shall not be required for metal veneer, except as is necessary to meet the fire-resistance requirements of the code.
     True   False

20. The area of a single section of thin exterior structural-glass veneer shall not exceed 32 square feet.
     True   False

21. The thickness of thin exterior structural-glass veneer shall be not less than 0.344 inch.
     True   False

22. Thin exterior structural-glass veneer shall be set only after the backing is moist.
     True   False

23. Where glass extends to a sidewalk surface, each section shall rest in an approved plastic molding.

    True   False

24. When glass extends to a sidewalk surface and is set in molding, the space between the molding and the sidewalk shall be thoroughly caulked and made watertight.

    True   False

25. MCM shall be permitted to be installed on buildings of Type V construction.

    True   False

MULTIPLE-CHOICE ANSWER KEY

1. B	9. C	17. D	25. D	33. C
2. A	10. A	18. C	26. C	34. D
3. C	11. A	19. C	27. C	35. A
4. C	12. D	20. D	28. C	36. D
5. C	13. C	21. C	29. B	37. B
6. C	14. C	22. C	30. C	38. A
7. C	15. C	23. C	31. C	39. D
8. B	16. B	24. D	32. C	40. A

TRUE-FALSE ANSWER KEY

1. T	6. F	11. T	16. T	21. T
2. T	7. T	12. T	17. T	22. F
3. F	8. F	13. T	18. F	23. F
4. F	9. T	14. T	19. T	24. T
5. T	10. F	15. F	20. F	25. T

Chapter 15

ROOF ASSEMBLIES AND ROOFTOP STRUCTURES

Welcome to Chapter 15 of the study guide. This chapter will test you on the provisions that govern the design, materials, construction, and quality of roof assemblies and rooftop structures. To succeed on this test, you must know the performance requirements, materials, and many other subjects. There have been changes in the code, so I cannot stress enough the importance of understanding as opposed to memorizing the code. Good luck.

MULTIPLE-CHOICE QUESTIONS

1. A layer of felt or nonbituminous saturated felt not less than _____ inches wide, shingled between each course of a wood-shake roof covering, is called interlayment.

    a. 10   b. 12

    c. 16   d. 18

2. A partially enclosed rooftop structure used to aesthetically conceal HVAC, electrical, or mechanical equipment from view is known as a(n) _____.

    a. equipment screen

    b. cover

    c. mechanical equipment screen

    d. none of the above

3. A metal roof panel is an interlocking metal sheet having a minimum installed weather exposure of ___ square foot/feet per sheet.

    a. 1   b. 2

    c. 3   d. 5

4. A metal roof shingle is an interlocking metal sheet having weather exposure of less than ___ square foot/feet per sheet.

    a. 1   b. 2

    c. 3   d. 5

5. A penthouse is an enclosed, unoccupied structure above the roof of a building, other than a _____, and not occupying more than one-third of the roof area.

    a. tank   b. basement

    c. both a and b   d. none of the above

6. A penthouse is an enclosed, unoccupied structure above the roof of a building, other than a _____, and not occupying more than one-third of the roof area.

    a. tower   b. basement

    c. both a and b   d. none of the above

7. A penthouse is an enclosed, unoccupied structure above the roof of a building, other than a _____, and not occupying more than one-third of the roof area.

    a. spire   b. dome cupola

    c. both a and b   d. none of the above

8. A penthouse is an enclosed, unoccupied structure above the roof of a building, other than a _____, and not occupying more than one-third of the roof area.

    a. spire   b. dome cupola

    c. bulkhead   d. all of the above

9. Positive roof drainage is a drainage condition in which consideration has been made for all loading deflections of the roof deck, and additional slope has been provided to ensure drainage of the roof within _____ hours of precipitation.

    a. 8   b. 12

    c. 24   d. 48

10. The process of installing an additional roof covering over a prepared existing roof covering without removing the existing roof covering is known as _____.

    a. roof paneling   b. roof replacement

    c. roof recovering   d. none of the above

11. Underlayment is one or more layers of _____.

    a. felt   b. sheathing paper

    c. either a or b   d. none of the above

!Codealert

See Section 1504.8 for changes in code requirements for gravel and stone.

> ## !Codealert
> A cricket or saddle shall be installed on the ridge side of any chimney or penetration greater than 30 inches wide as measured perpendicular to the slope. Cricket or saddle coverings shall be of sheet metal or of the same material as the roof covering.

12. Roof assemblies shall be divided into classes, which may include _____.

    a. A

    b. B

    c. C

    d. all of the above

13. Nonclassified roofing is approved material that is not listed as a Class _____ roof covering.

    a. A

    b. B

    c. C

    d. none of the above

14. Fasteners for asphalt shingles shall be nails with a _____ finish.

    a. galvanized

    b. stainless-steel

    c. either a or b

    d. none of the above

15. Fasteners for asphalt shingles shall be nails with a _____ finish.

    a. galvanized

    b. stainless-steel

    c. copper

    d. all of the above

16. Asphalt shingles shall be secured to a roof with not less than _____ fasteners per strip shingle or two fasteners per individual shingle.

    a. three

    b. four

    c. five

    d. six

17. Underlayment application requires two layers when the roof slope ranges from 17 percent to _____ percent.

    a. 22

    b. 28

    c. 30

    d. 33

18. When an ice-dam membrane is required, the membrane must extend from the eave's edge to a point at least _____ inches inside the exterior wall line of the building.

    a. 12   b. 16

    c. 18   d. 24

19. Clay and concrete roof tile shall be installed on roof slopes with a _____-percent or greater slope.

    a. 18   b. 20

    c. 21   d. 25

20. Roofs with a minimum slope of 33 percent require underlayment to be installed in no less than _____.

    a. one layer   b. two layers

    c. either a or b   d. none of the above

21. When used as a valley lining material, aluminum must have a minimum thickness of _____ inch.

    a. 0.010   b. 0.024

    c. 0.026   d. 0.028

22. When used as a valley lining material, copper must have a minimum weight of _____ ounces.

    a. 12   b. 14

    c. 16   d. 24

23. Tile fasteners shall be corrosion-resistant and not less than _____ -gauge.

    a. 8   b. 10

    c. 11   d. 12

24. The minimum slope for lapped, nonsoldered seam metal roofs without applied lap sealant shall be _____ percent.

    a. 17   b. 20

    c. 25   d. 33

25. The minimum slope for lapped, nonsoldered seam metal roofs with applied lap sealant shall be _____ percent.

    a. 4   b. 10

    c. 15   d. 20

26. Metal roof shingles shall not be installed on roof slopes below _____ percent.

    a. 17   b. 20

    c. 25   d. 33

27. Mineral-surfaced roll roofing shall not be applied on roof slopes with a slope less than _____ percent.

    a. 6   b. 8

    c. 17   d. 22

28. Slate shingles shall only be used on slopes of _____ or greater.

    a. 4:12   b. 4:15

    c. 5:20   d. 5:22

29. Flashing and counterflashing shall be made with _____.

    a. asphalt shingles   b. sheet metal

    c. wood   d. any of the above

30. Wood shingles shall be installed on roofs with a slope of _____ percent or greater.

    a. 17   b. 22

    c. 25   d. 33

31. Fasteners for wood shingles shall be corrosion-resistant with a minimum penetration of _____ inch into the sheathing.

    a. 1/4   b. 1/3

    c. 1/2   d. 3/4

32. Wood shingles shall be laid with a side lap not less than _____ inches between joints in adjacent courses.

    a. 1 1/4   b. 1 1/2

    c. 1 3/4   d. 2

!**Code**alert

Code requirements for material standards have changed. See Section 1508.2 for the updates.

33. Solid sheathing is required in areas where the average daily temperature in January is _____ degrees F or less or where there is a possibility of ice forming along the eaves, causing a backup of water.

    a. 0   b. 15

    c. 25   d. 32

34. Thermoset single-ply membrane roofs shall be designed at a minimum of a ____-percent slope for drainage.

    a. 1   b. 2

    c. 5   d. 10

35. Sprayed polyurethane-foam roofs shall be designed at a minimum of a ____-percent slope for drainage.

    a. 1   b. 2

    c. 5   d. 10

TRUE-FALSE QUESTIONS

1. Built-up roofing consists of two or more layers of felt cemented together and surfaced with a cap sheet, mineral aggregate, smooth coating, or similar surfacing material.

    True   False

2. Modified bitumen roof covering is one or more layers of polymer-modified asphalt sheets.

    True   False

3.   A roof assembly is a system designed to provide weather protection and resistance to design loads.

      True   False

4.   A roof deck consists of the flat or sloped surface, including its support members or vertical supports.

      True   False

5.   Roof ventilation can result from a natural or a mechanical process.

      True   False

6.   An opening in a wall or parapet that allows water to drain from a roof is called a roof leader.

      True   False

7.   Roof flashings are designed to prevent moisture from entering a wall or roof though joints in copings.

      True   False

8.   Flashing is required to be installed at wall and roof intersections and at gutters.

      True   False

9.   All gutters for all uses must be made of noncombustible materials.

      True   False

10.   Parapet walls shall be properly coped with noncombustible, weatherproof materials of a width no less than the thickness of the parapet wall.

      True   False

11.   Gravel or stone shall not be used on the roof of a building located in a hurricane-prone region.

      True   False

!Codealert

When reroofing, the application of a new protective coating over an existing spray polyurethane-foam roofing system shall be permitted without tear-off of existing roof coverings.

12. Class B roof assemblies are those that are effective against severe fire-test exposure.

True   False

13. Class A roof assemblies shall be permitted for use in buildings or structures of all types of construction.

True   False

14. Class A roof assemblies are those that are effective against moderate fire-test exposure.

True   False

15. Class C roof assemblies are those that are effective against light fire-test exposure.

True   False

16. Fire-retardant-treated wood shakes and shingles shall be treated by impregnation with chemicals by the full-cell vacuum-pressure process.

True   False

17. Asphalt shingles shall be fastened to solidly sheathed decks.

True   False

18. Asphalt shingles cannot be installed on a roof that has a slope of less than 20 percent.

True   False

19. Drip edge is required at the eaves and gables of shingle roofs.

True   False

20. Eave drip edges shall extend a minimum of 1/2 inch below sheathing and back on the roof a minimum of 2 inches.

True   False

21. Drip edge shall be mechanically fastened a maximum of 16 inches on center.

True   False

22. A cricket or saddle shall be installed on the ridge side of any chimney or penetration greater than 24 inches wide as measured perpendicular to the slope.

True   False

23. Cricket and saddle coverings must be made of the same material as the roof covering.

True   False

24. Clay tile shall be installed only over solid sheathing or spaced structural sheathing boards.
    True   False

25. Concrete roof tile may not be installed on roofs that have a slope of 23 percent or less.
    True   False

MULTIPLE-CHOICE ANSWER KEY

1. D	8. D	15. D	22. C	29. B
2. C	9. D	16. B	23. C	30. C
3. C	10. C	17. C	24. C	31. D
4. C	11. C	18. D	25. A	32. B
5. A	12. D	19. C	26. C	33. C
6. A	13. D	20. A	27. B	34. B
7. C	14. C	21. B	28. A	35. B

TRUE-FALSE ANSWER KEY

1. T	6. F	11. T	16. T	21. F
2. T	7. T	12. F	17. T	22. F
3. T	8. T	13. T	18. F	23. F
4. F	9. F	14. F	19. T	24. T
5. T	10. T	15. T	20. F	25. F

Chapter 16

STRUCTURAL DESIGN

Welcome to the study-guide test for structural design. If you know the code, you will not have a problem passing this sample test. It is designed to help you prepare for taking the real test. This chapter contains questions regarding the structural design of buildings, structures, and portions therein regulated by this code. You must know the types of paperwork you will need to include in your construction documents, the difference between loads, and many other subjects. This study guide will help you to learn the code in terms that you can understand. On with the test...

MULTIPLE-CHOICE QUESTIONS

1. An exterior floor projecting from and supported by a structure without additional independent supports is called _____.

    a. a deck
    b. an exterior balcony
    c. a balcony
    d. both b and c

2. An exterior floor supported on at least two opposing sides by an adjacent structure and/or posts, piers, or other independent supports is known as _____.

    a. a deck
    b. an exterior balcony
    c. a balcony
    d. both b and c

3. The product of the nominal strength and a resistance factor is known as _____.

    a. maximum strength
    b. sensible strength
    c. design strength
    d. none of the above

4. A horizontal or sloped system acting to transmit lateral forces to the vertical-resisting elements is known as _____.

    a. a header
    b. a lintel
    b. a diaphragm
    d. none of the above

5. Essential facilities are building and other structures that are intended to remain operational in the event of extreme environmental loading from _____.

    a. snow
    b. sand
    c. ice
    d. all of the above

6. Essential facilities are building and other structures that are intended to remain operational in the event of extreme environmental loading from _____.

    a. rubble   b. sand

    c. flood   d. all of the above

7. Essential facilities are building and other structures that are intended to remain operational in the event of extreme environmental loading from _____.

    a. earthquakes   b. wind

    c. flood   d. all of the above

8. Fabric partitions consist of a finished surface made of fabric without a continuous rigid backing that is directly attached to a framing system in which the vertical framing members are spaced greater than _____ feet on center.

    a. 2   b. 3

    c. 4   d. 5

9. A condition beyond which a structure or member becomes unfit for service and is judged to be no longer useful for its intended function or unsafe is known as _____.

    a. risky   b. stable

    c. limit state   d. condemned

10. Loads produced by the use and occupancy of a building or other structure and that do not include construction or environmental loads such as wind, snow, rain, earthquake, flood, or dead load are known as _____.

    a. live loads   b. human loads

    c. substantial loads   d. none of the above

!Codealert

A fabric partition consisting of a finished surface made of fabric, without a continuous rigid backing, that is directly attached to a framing system in which the vertical framing members are spaced greater than 4 feet on center.

!Codealert

Occupancy category is a category used to determine structural requirements.

11. A factor that accounts for deviations of the actual load from the nominal load, for uncertainties in the analysis that transforms the load into a load effect, and for the probability that more than one extreme load will occur simultaneously is known as _____.

   a. a live load   b. a load effect

   c. a load factor   d. either b or c

12. Forces or other actions that result from the weight of building materials, occupants and their possessions, environmental effects, differential movement, and restrained dimensional changes are known as _____.

   a. free forces   b. loads

   c. pull   d. none of the above

13. Nominal loads refer to _____ loads.

   a. wind   b. snow

   c. both a and b   d. none of the above

14. Nominal loads refer to _____ loads.

   a. rain   b. atmospheric

   c. both a and b   d. none of the above

15. Nominal loads refer to _____ loads.

   a. soil   b. live

   c. both a and b   d. none of the above

16. Nominal loads refer to _____ loads.

   a. negligible   b. full

   c. both a and b   d. none of the above

17. Nominal loads refer to _____ loads.

    a. flood   b. full

    c. both a and b   d. none of the above

18. Nominal loads refer to _____ loads.

    a. flood   b. earthquake

    c. both a and b   d. none of the above

19. The section of a floor, wall, or roof between the supporting frame of two adjacent rows of columns, girders, or column bands of floor or roof construction is known as a(n)_____.

    a. invert   b. panel

    c. co-element   d. none of the above

20. Resistance factor is a factor that accounts for deviations of the _____ strength from the nominal strength and the manner and consequences of failure.

    a. normal   b. nominal

    c. maximum   d. actual

21. A system of building components near open sides of a garage floor or ramp or a system of building walls that act as a restraint for vehicles is known as a _____.

    a. roadblock system   b. rail system

    c. vehicle barrier system   d. none of the above

22. Construction documents shall show the_____ of structural members with floor levels, column centers, and offsets dimensioned.

    a. size   b. section

    c. relative locations   d. all of the above

23. Design requirements include such factors as _____.

    a. strength design   b. load

    c. both a and b   d. none of the above

24. Design requirements include such factors as _____.

    a. resistance-factor design   b. load

    c. both a and b   d. none of the above

25. Design requirements include such factors as _____.
    a. strength design
    b. overtime expenses
    c. both a and b
    d. none of the above

26. Design requirements include such factors as _____.
    a. allowable stress design
    b. refractor design
    c. both a and b
    d. none of the above

27. Design requirements include such factors as _____.
    a. allowable stress design
    b. empirical design
    c. both a and b
    d. none of the above

28. Design requirements include such factors as _____.
    a. conventional construction methods
    b. empirical design
    c. both a and b
    d. none of the above

29. Structural systems and members thereof shall be designed to have adequate _____ to limit deflections and lateral drift.
    a. stiffness
    b. flexibility
    c. both a and b
    d. none of the above

30. Load effects on structural members and their connections shall be determined by methods of structural analysis that take into account _____.
    a. equilibrium
    b. licensing
    c. both a and b
    d. none of the above

31. Load effects on structural members and their connections shall be determined by methods of structural analysis that take into account _____.
    a. equilibrium
    b. general stability
    c. both a and b
    d. none of the above

> **!Code**alert
>
> A system of building components near open sides of a garage floor or ramp or building walls that act as restraints for vehicles is a vehicle barrier system.

32. Load effects on structural members and their connections shall be determined by methods of structural analysis that take into account _____.

    a. equilibrium
    b. geometric compatibility
    c. both a and b
    d. none of the above

33. Load effects on structural members and their connections shall be determined by methods of structural analysis that take into account _____.

    a. short-term material properties
    b. geometric compatibility
    c. long-term material properties
    d. all of the above

34. Concrete and masonry walls are required to be anchored to provide lateral support. Suitable anchorage locations include _____.

    a. floors
    b. roofs
    c. structural elements
    d. all of the above

35. Where supported by attachment to an exterior wall, decks shall be positively anchored to the primary structure and designed for _____ loads as applicable.

    a. vertical
    b. lateral
    c. both a and b
    d. none of the above

36. In office buildings and in other buildings where partition locations are subject to change, provisions for partition weight shall be made, whether or not partitions are shown on the construction documents, unless the specified live load exceeds _____ psf.

    a. 40
    b. 60
    c. 80
    d. 100

37. The concentrated and uniform loads shall be uniformly distributed over a _____-foot width on a line normal to the centerline out placed within a 12 foot-wide lane.

    a. 8   b. 10

    c. 12   d. none of the above

38. Handrail assemblies and guards shall be designed to resist a load of _____ psf applied in any direction at the top and to transfer this load through the supports to the structure.

    a. 20   b. 25

    c. 45   d. 50

39. Intermediate rails (those that are not handrails), such as balusters and panel fillers, shall be designed to withstand a horizontally applied normal load of _____ pounds on an area equal to 1 square foot, including openings and space between rails.

    a. 20   b. 25

    c. 45   d. 50

40. Vehicle barrier systems for passenger cars shall be designed to resist a single load of _____ pounds applied horizontally in any direction to the barrier system and shall have anchorage or attachment capable of transmitting this load to the structure.

    a. 2,000   b. 4,000

    c. 6,000   d. 10,000

41. Special-purpose roofs are roofs that are designed to be used for _____.

    a. promenade purposes   b. roof gardens

    c. assembly purposes   d. all of the above

42. The maximum wheel load of a crane, when figuring vertical impact force for a powered

!Codealert

This chapter contains many code changes, additions, and updates. Consult your code book carefully to become aware of all the new code language.

monorail crane, shall be increased by _____ percent.

a. 15   b. 20

c. 25   d. 50

43. A flood with a 1-percent chance of being equaled or exceeded in any given year is a _____.

a. nominal flood   b. normal flood

c. base flood   d. none of the above

44. The portion of a building with its floor subgrade on all sides is known as the _____.

a. cellar   b. basement

c. crawlspace   d. either a or b

45. A combination of design modifications that results in a building or structure, including the attendant utility and sanitary facilities, being watertight with walls substantially impermeable to the passage of water and with structural components having the capacity to resist loads is called _____.

a. waterproofing   b. purging

c. dry floodproofing   d. none of the above

46. A flood can be a general and temporary condition or a partial or complete inundation of normally dry land from the overflow of _____ waters.

a. tidal   b. inland

c. either a or b   d. none of the above

47. A flood can be a general and temporary condition or a partial or complete inundation of normally dry land from the unusual and rapid accumulation of _____ waters from any source.

a. runoff   b. surface

c. either a or b   d. none of the above

48. When dealing with earthquake loads, a condition of being in two horizontal directions at 90 degrees to each other is known as _____.

a. orthogonal   b. ornamental

c. passive   d. none of the above

49. When dealing with site-class definitions, the soil-profile name is hard rock and the site class is _____.

    a. A   b. B

    c. C   d. D

50. When dealing with site-class definitions, the soil-profile name is rock and the site class is _____.

    a. A   b. B

    c. C   d. D

TRUE-FALSE QUESTIONS

1. Allowable stress design is a method of proportioning structural members such that elastically computed stresses produced in the members by nominal loads do not exceed specified allowable stresses.

    True   False

2. A blocked diaphragm is found in light-frame construction.

    True   False

3. A flexible diaphragm bends in order to distribute story shear and torsional moment.

    True   False

4. The period of continuous application of a given load or the aggregate of periods of intermittent applications of the same load is known as a continuous load.

    True   False

5. The product of a nominal load and a load factor is a fractional load.

    True   False

6. An impact load is one that results from moving machinery.

    True   False

7. An impact load is one that results from elevators.

    True   False

8. Roof live loads are loads produced during maintenance by workers, equipment, and material and during the life of the structure by movable objects, such as planters and people.

    True   False

9. Forces and deformations produced in structural members by applied loads are known as load factors.

    True   False

10. A category used to determine structural requirements based on occupancy is known as the occupancy category.

    True   False

11. Required strength is the strength of a member, cross section, or connection required to assist factored loads.

    True   False

12. Nominal strength is the capacity of a structure or member to resist the effects of loads as determined by computations using specified material strengths and dimensions, equations derived from accepted principles of structural mechanics, or by field or laboratory tests of scaled models, allowing for modeling effects and differences between laboratory and field conditions.

    True   False

13. LRFD is an acronym for load and regular factor design.

    True   False

14. It is unlawful to remove or deface notices of posted live loads.

    True   False

15. It is unlawful to place, cause, or permit to be placed on any floor, roof, building, structure, or portion thereof a load greater than is permitted by local regulations.

    True   False

16. Structures shall be designed and constructed in accordance with ADFC regulations.

    True   False

17. Loads and forces for occupancies or uses not covered in the code book are subject to the approval of the municipal study group.

    True   False

18. All buildings shall be assigned an occupancy category.

   True   False

19. The anchorage of a roof to walls and columns shall be provided to resist the uplift and sliding forces that result from the application of the prescribed loads.

   True   False

20. Concrete and masonry walls shall be anchored to floors to provide vertical support.

   True   False

21. Dead loads shall not be considered permanent loads.

   True   False

22. When working with dead loads for the purposes of design, the actual weights of materials of construction and fixed service equipment shall be used.

   True   False

23. The live loads used in the design of buildings and other structures shall be the maximum loads expected by the intended use or occupancy but shall in no case be less than the minimum uniformly distributed unit loads.

   True   False

24. The minimum live loads for garages having trucks or buses shall not be less than 65 psf.

   True   False

25. Handrail assemblies and guards shall be able to resist a single concentrated load of 150 pounds.

   True   False

26. Grab bars shall be designed to resist a single concentrated load of 250 pounds applied in any direction at any point.

   True   False

27. Shower seats shall be designed to resist a single concentrated load of 275 pounds applied in any direction at any point.

   True   False

28. Dressing-room bench seat systems shall be designed to resist a single concentrated load of 250 pounds applied in any direction at any point.

   True   False

29. Elevator loads shall be increased by 50 percent for impact and structural supports.

True   False

30. Live loads of 100 psf or less shall not be reduced in public assembly occupancies.

True   False

31. With the proper design methods roofs may be landscaped.

True   False

32. A crane live load shall be the rated capacity of the crane.

True   False

33. In wind-borne-debris regions, glazing in buildings shall be impact-resistant or protected with an impact-resistant covering meeting the requirements of an approved impact-resisting standard.

True   False

34. Basement walls shall be designed to resist lateral soil loads.

True   False

35. Roofs equipped with hardware to control the rate of drainage shall be equipped with a secondary drainage system at a higher elevation that limits accumulation of water on the roof above the elevation.

True   False

36. Flood-damage-resistant materials consist of materials that are capable of withstanding direct and prolonged contact with floodwaters without sustaining any damage that requires more than cosmetic repair.

True   False

37. FIRM stands for flood-insurance-rate map.

True   False

38. The earthquake ground motion that buildings and structures are specifically proportioned to resist is known as nominal earthquake action.

True   False

39. The most severe earthquake effects considered by the code are known as maximum considered earthquake ground motion.

True   False

40. When dealing with earthquake loads, mechanical systems shall not consider plumbing systems as a part of the mechanical system.

   True   False

FIGURE 16.1   Ground snow loads, P*g*, for the United States (psf).

FIGURE 16.1   Ground snow loads, P*g*, for the United States (psf).   *(continued)*

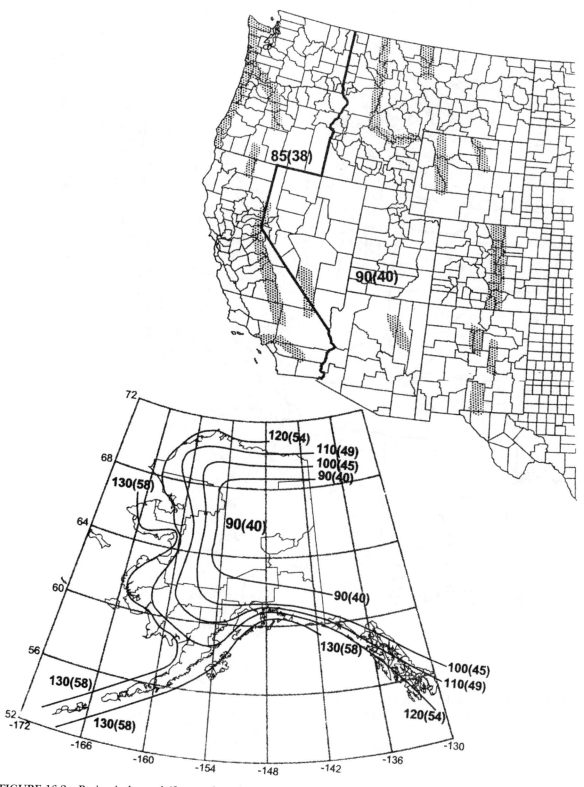

FIGURE 16.2   Basic wind speed (3-second gust).

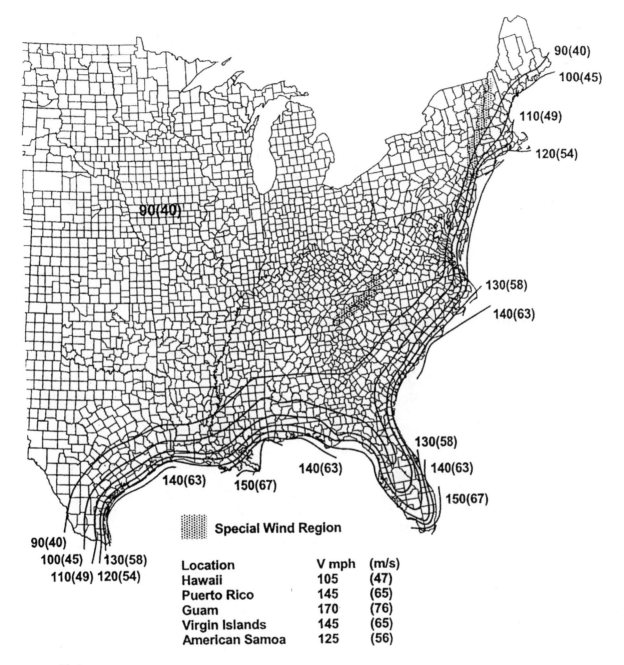

Location	V mph	(m/s)
Hawaii	105	(47)
Puerto Rico	145	(65)
Guam	170	(76)
Virgin Islands	145	(65)
American Samoa	125	(56)

Notes:
1. Values are nominal design 3-second gust wind speeds in miles per hour (m/s) at 33 ft (10 m) above ground for Exposure C category.
2. Linear interpolation between wind contours is permitted.
3. Islands and coastal areas outside the last contour shall use the last wind speed contour of the coastal area.
4. Mountainous terrain, gorges, ocean promontories, and special wind regions shall be examined for unusual wind conditions.

FIGURE 16.2   Basic wind speed (3-second gust). *(continued)*

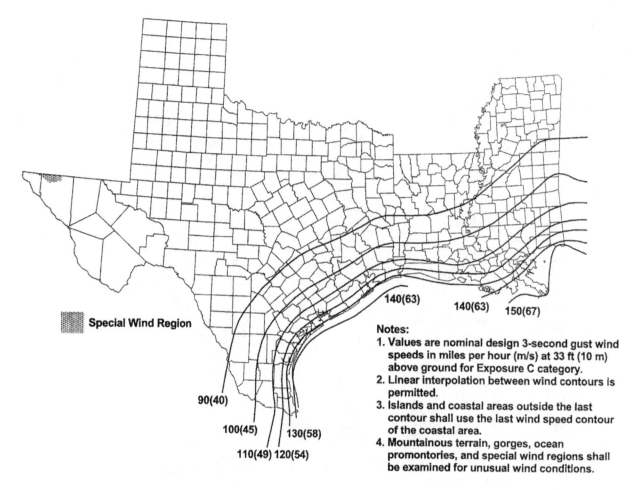

FIGURE 16.2 Basic wind speed (3-second gust), western Gulf of Mexico hurricane coastline. *(continued)*

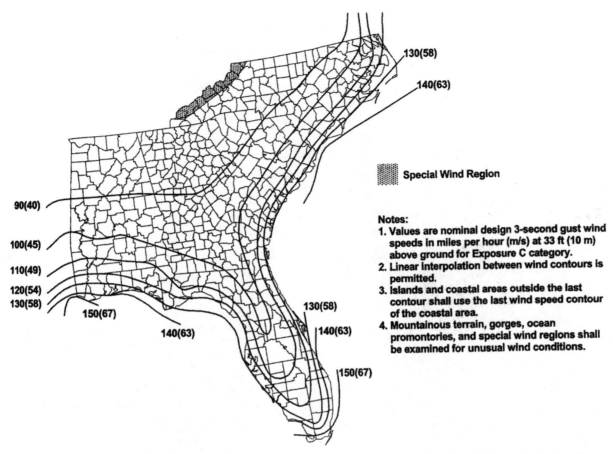

130(58)

140(63)

Special Wind Region

90(40)

100(45)

110(49)

120(54)
130(58)

150(67)

140(63)

130(58)

140(63)

150(67)

Notes:
1. Values are nominal design 3-second gust wind speeds in miles per hour (m/s) at 33 ft (10 m) above ground for Exposure C category.
2. Linear interpolation between wind contours is permitted.
3. Islands and coastal areas outside the last contour shall use the last wind speed contour of the coastal area.
4. Mountainous terrain, gorges, ocean promontories, and special wind regions shall be examined for unusual wind conditions.

FIGURE 16.2 Basic wind speed (3-second gust), eastern Gulf of Mexico and southeastern U.S. hurricane coastline. *(continued)*

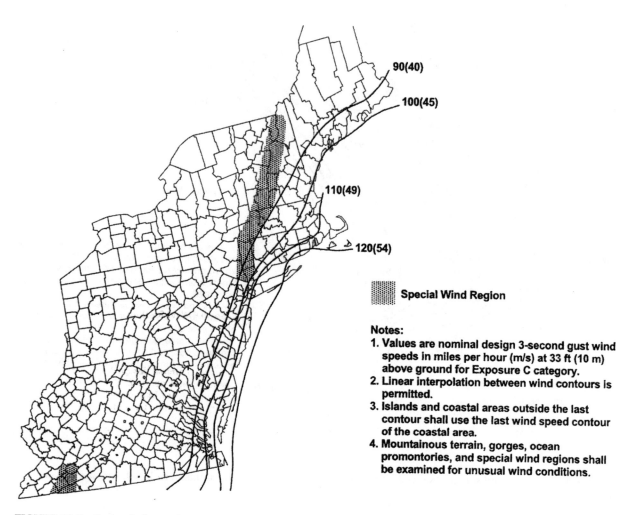

Special Wind Region

Notes:
1. Values are nominal design 3-second gust wind speeds in miles per hour (m/s) at 33 ft (10 m) above ground for Exposure C category.
2. Linear interpolation between wind contours is permitted.
3. Islands and coastal areas outside the last contour shall use the last wind speed contour of the coastal area.
4. Mountainous terrain, gorges, ocean promontories, and special wind regions shall be examined for unusual wind conditions.

FIGURE 16.2 Basic wind speed (3-second gust), mid and northern Atlantic hurricane coastline. *(continued)*

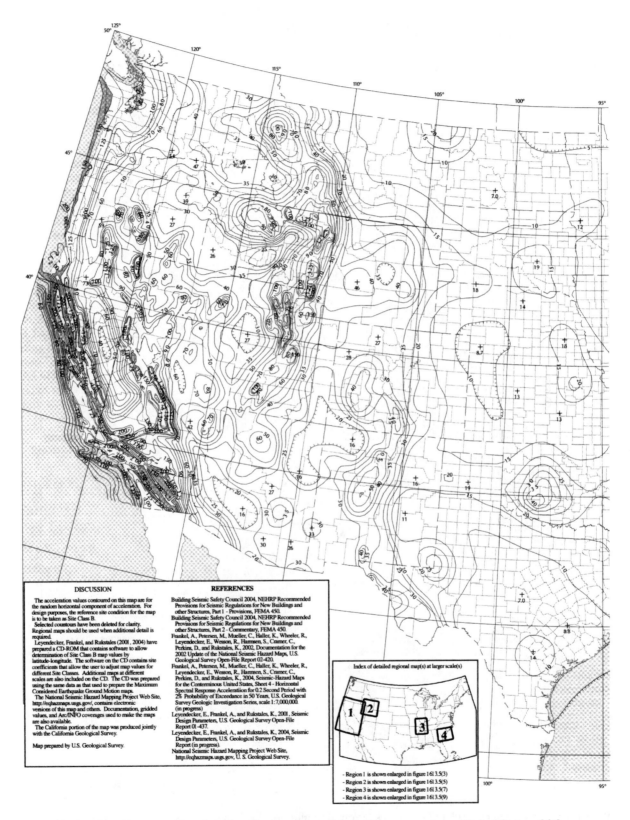

FIGURE 16.3 Maximum considered earthquake ground motion for the conterminous United States of 0.2 sec spectral response acceleration (5% of critical damping), Site Class B.

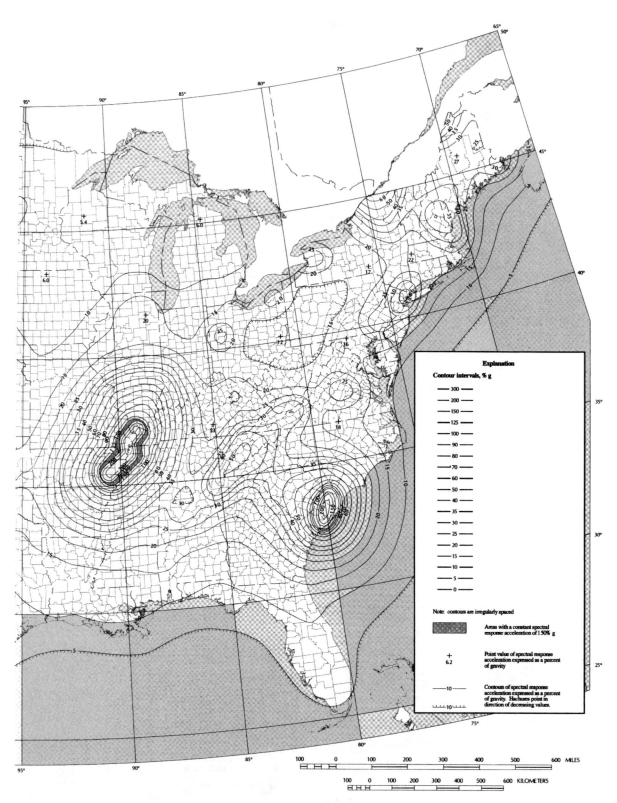

FIGURE 16.3   Maximum considered earthquake ground motion for the conterminous United States of 0.2 sec spectral response acceleration (5% of critical damping), Site Class B. *(continued)*

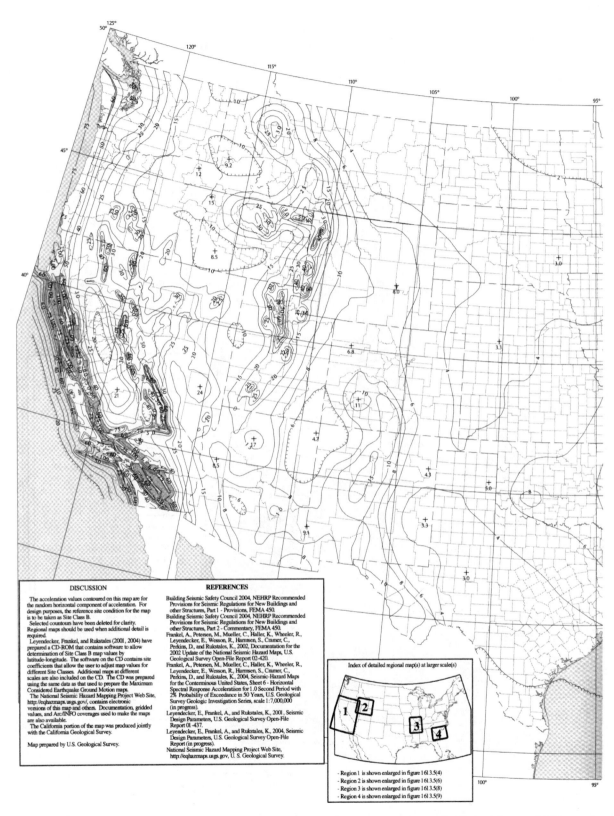

REFERENCES

Building Seismic Safety Council 2004, NEHRP Recommended Provisions for Seismic Regulations for New Buildings and other Structures, Part 1 - Provisions, FEMA 450.

Building Seismic Safety Council 2004, NEHRP Recommended Provisions for Seismic Regulations for New Buildings and other Structures, Part 2 - Commentary, FEMA 450.

Frankel, A, Petersen, M., Mueller, C., Haller, K., Wheeler, R., Leyendecker, E., Wesson, R., Harmsen, S., Cramer, C., Perkins, D., and Rukstales, K., 2002, Documentation for the 2002 Update of the National Seismic Hazard Maps, U.S. Geological Survey Open-File Report 02-420.

Frankel, A., Petersen, M., Mueller, C., Haller, K., Wheeler, R., Leyendecker, E., Wesson, R., Harmsen, S., Cramer, C., Perkins, D., and Rukstales, K., 2004, Seismic-Hazard Maps for the Conterminous United States, Sheet 6 - Horizontal Spectral Response Acceleration for 1.0 Second Period with 2% Probability of Exceedance in 50 Years, U.S. Geological Survey Geologic Investigation Series, scale 1:7,000,000 (in progress).

Leyendecker, E., Frankel, A., and Rukstales, K., 2001, Seismic Design Parameters, U.S. Geological Survey Open-File Report 01-437.

Leyendecker, E., Frankel, A., and Rukstales, K., 2004, Seismic Design Parameters, U.S. Geological Survey Open-File Report (in progress).

National Seismic Hazard Mapping Project Web Site, http://eqhazmaps.usgs.gov, U. S. Geological Survey.

Index of detailed regional map(s) at larger scale(s)

- Region 1 is shown enlarged in figure 16l 3.5(4)
- Region 2 is shown enlarged in figure 16l 3.5(6)
- Region 3 is shown enlarged in figure 16l 3.5(8)
- Region 4 is shown enlarged in figure 16l 3.5(9)

FIGURE 16.4 Maximum considered earthquake ground motion for the conterminous United States of 1.0 sec spectral response acceleration (5% of critical damping), Site Class B.

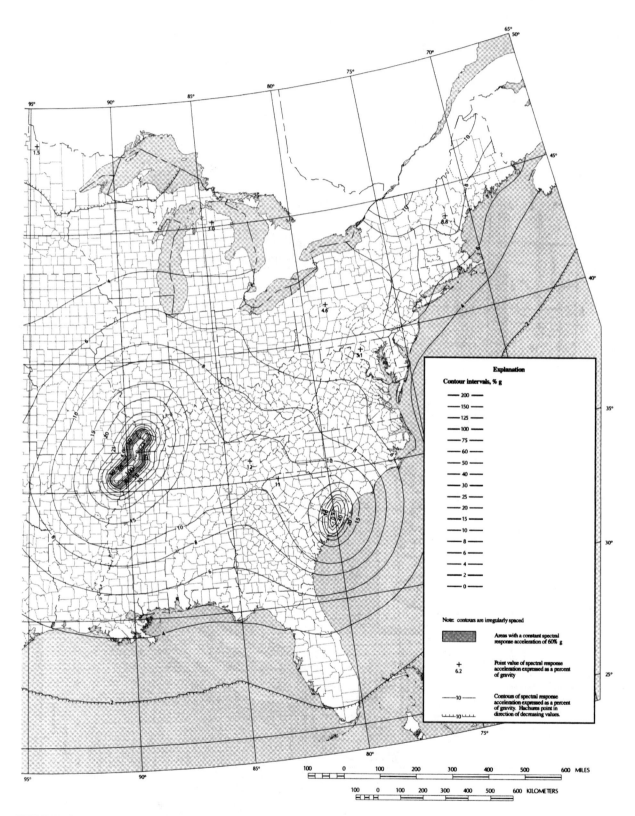

FIGURE 16.4   Maximum considered earthquake ground motion for the conterminous United States of 1.0 sec spectral response acceleration (5% of critical damping), Site Class B. *(continued)*

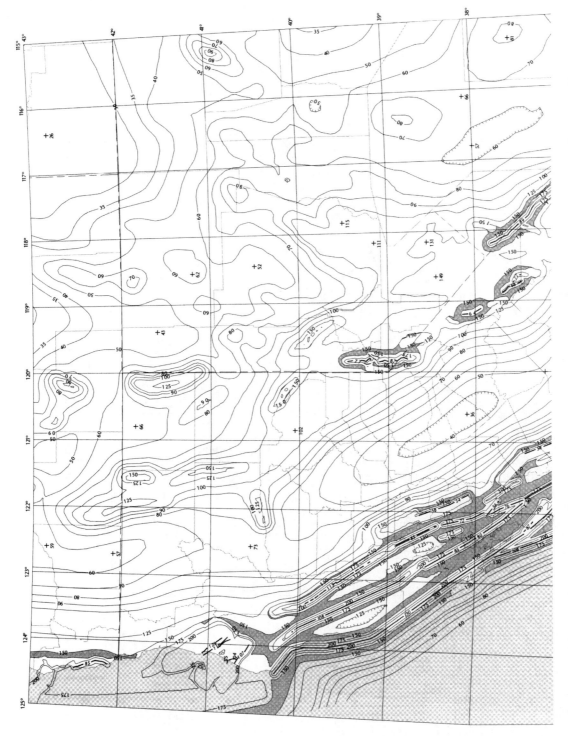

FIGURE 16.5 Maximum considered earthquake ground motion for Region 1 of 0.2 sec spectral response acceleration (5% of critical damping), Site Class B.

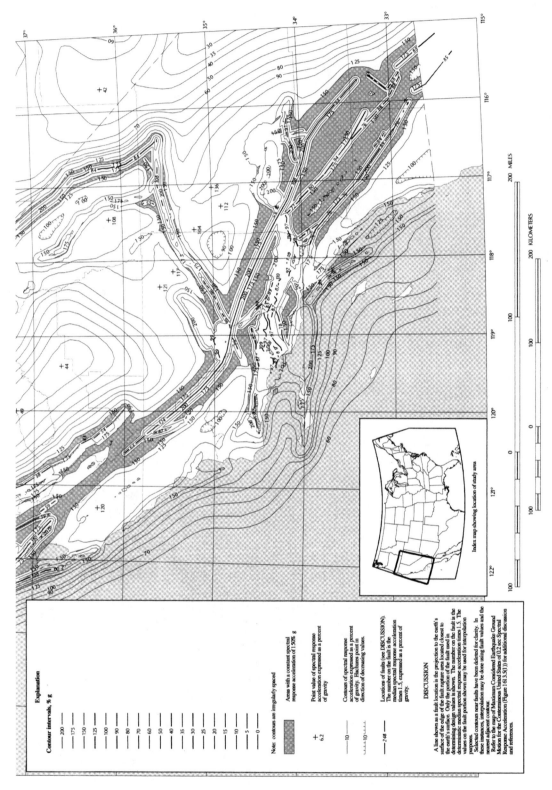

FIGURE 16.5   Maximum considered earthquake ground motion for Region 1 of 0.2 sec spectral response acceleration (5% of critical damping), Site Class B. *(continued)*

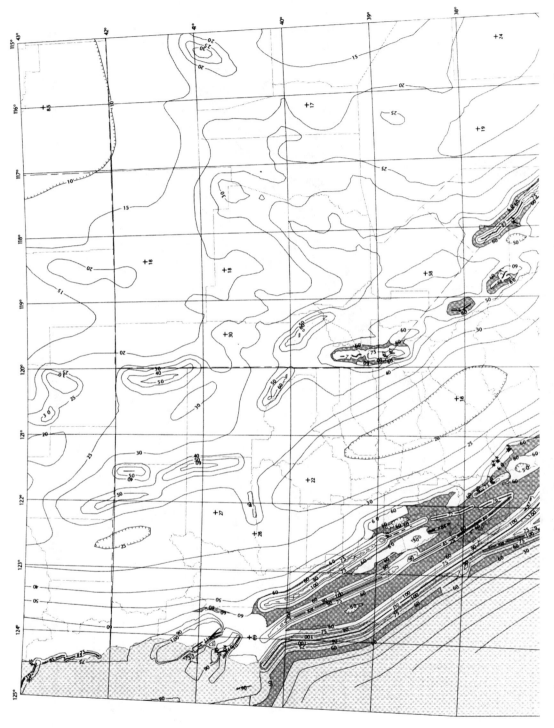

FIGURE 16.6 Maximum considered earthquake ground motion for Region1 of 1.0 sec spectral response acceleration (5% of critical damping), Site Class B.

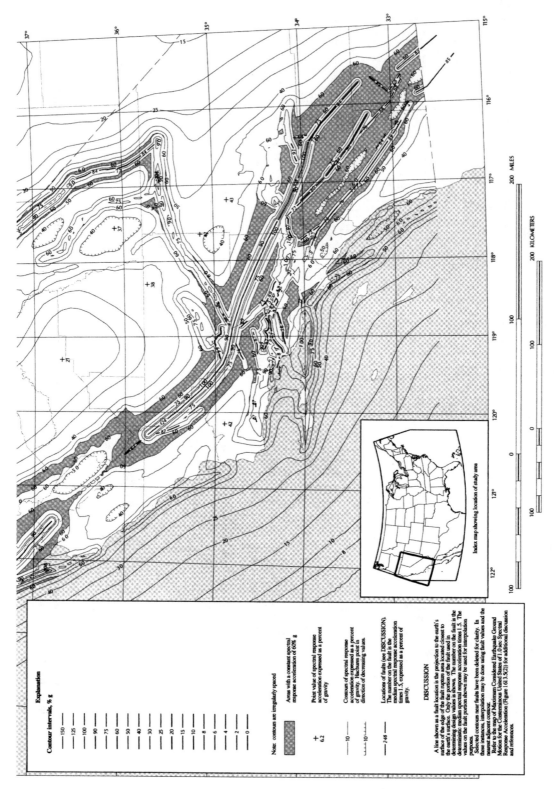

FIGURE 16.6   Maximum considered earthquake ground motion for Region1 of 1.0 sec spectral response acceleration (5% of critical damping), Site Class B. *(continued)*

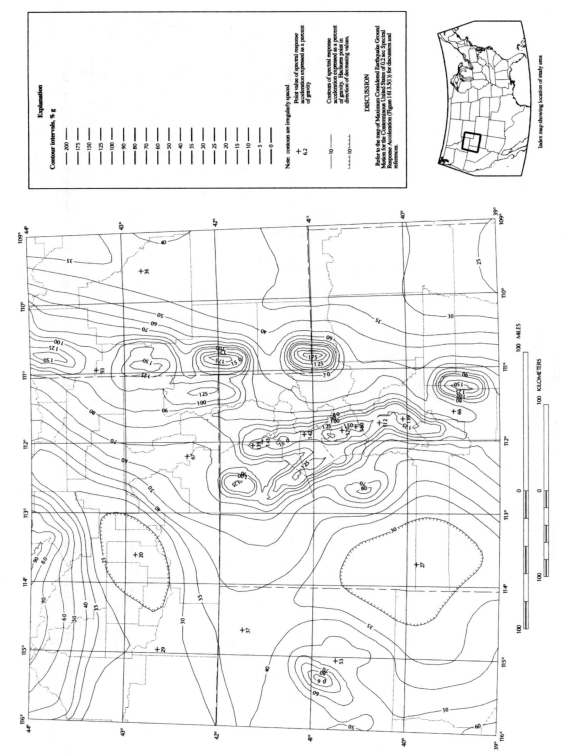

FIGURE 16.7 Maximum considered earthquake ground motion for Region 2 of 0.2 sec spectral response acceleration (5% of critical damping), Site Class B.

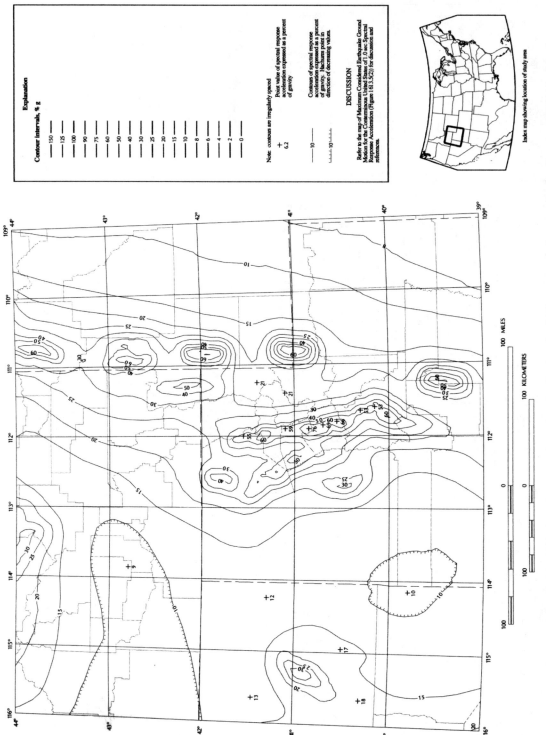

FIGURE 16.8   Maximum considered earthquake ground motion for Region 2 of 1.0 sec spectral response acceleration (5% of critical damping), Site Class B.

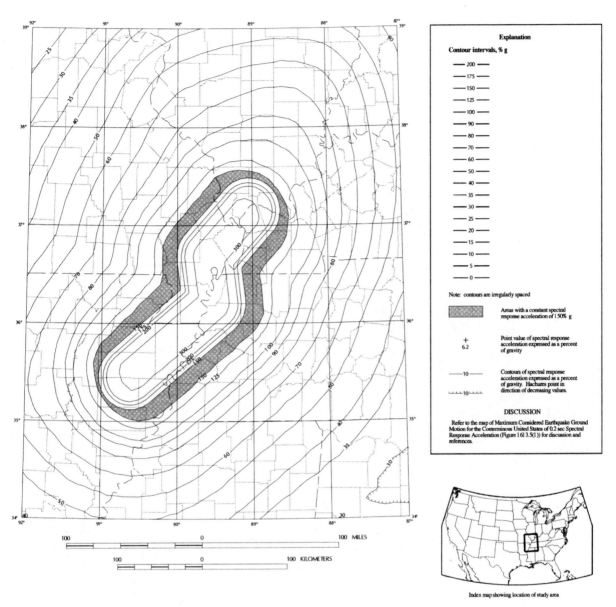

FIGURE 16.9   Maximum considered earthquake ground motion for Region 3 of 0.2 sec spectral response acceleration (5% of critical damping), Site Class B.

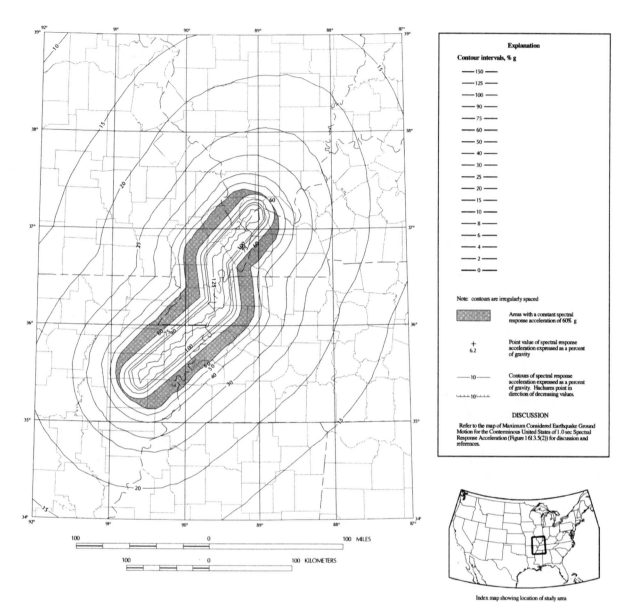

FIGURE 16.10   Maximum considered earthquake ground motion for Region 3 of 1.0 sec spectral response acceleration (5% of critical damping), Site Class B.

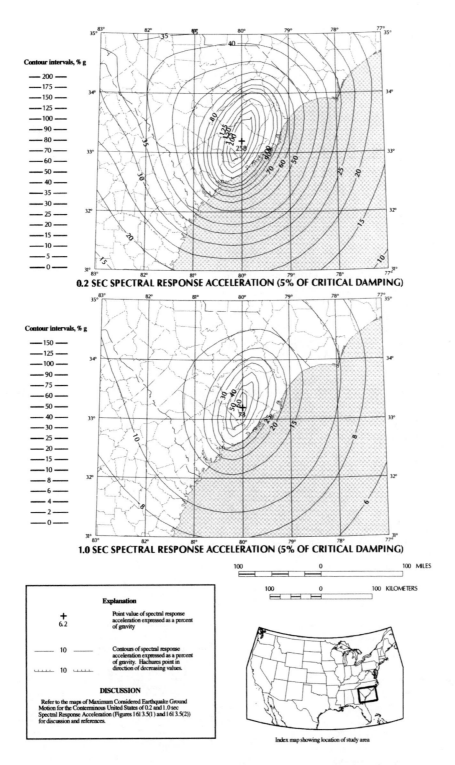

FIGURE 16.11 Maximum considered earthquake ground motion for Region 4 of 0.2 and 1.0 sec spectral response acceleration (5% of critical damping), Site Class B.

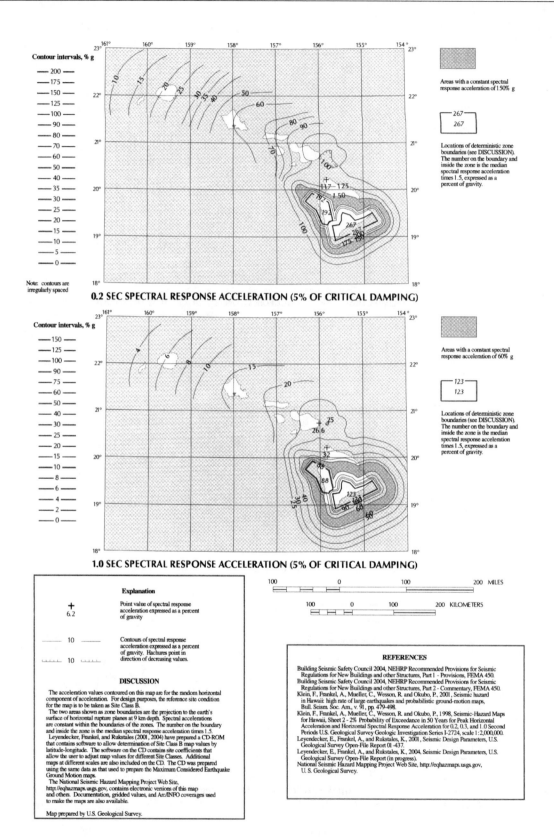

FIGURE 16.12   Maximum considered earthquake ground motion for Hawaii of 0.2 and 1.0 sec spectral response acceleration (5% of critical damping), Site Class B.

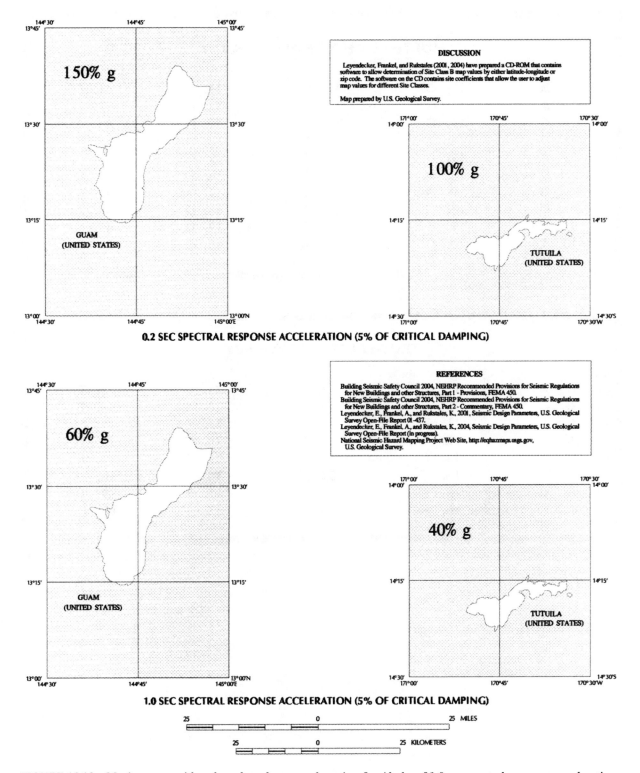

FIGURE 16.13 Maximum considered earthquake ground motion for Alaska of 1.0 sec spectral response acceleration (5% of critical damping), Site Class B.

MULTIPLE-CHOICE ANSWER KEY

1. B	11. C	21. C	31. C	41. D
2. A	12. B	22. D	32. C	42. C
3. C	13. C	23. C	33. D	43. C
4. B	14. A	24. C	34. D	44. B
5. A	15. C	25. A	35. C	45. C
6. C	16. D	26. A	36. C	46. C
7. D	17. A	27. C	37. B	47. C
8. C	18. C	28. C	38. D	48. A
9. C	19. B	29. A	39. D	49. A
10. A	20. D	30. A	40. C	50. B

TRUE-FALSE ANSWER KEY

1. T	9. F	17. F	25. F	33. T
2. T	10. T	18. T	26. T	34. T
3. T	11. F	19. T	27. F	35. F
4. F	12. T	20. F	28. T	36. T
5. F	13. F	21. F	29. F	37. T
6. T	14. T	22. T	30. T	38. F
7. T	15. T	23. T	31. T	39. T
8. T	16. F	24. F	32. T	40. F

Chapter 17

STRUCTURAL TESTS AND SPECIAL INSPECTIONS

You are about to take the sample test for Chapter 17. All materials of construction and tests must conform to the applicable standards listed in this code, but are you comfortable in your knowledge regarding these materials? Are you ready to take this sample test? There were code changes for 2006, so remember that it is your responsibility to be aware of them. This is the perfect opportunity to test your knowledge.

MULTIPLE-CHOICE QUESTIONS

1. An assemblage of structural elements assigned to provide support and stability for an overall structure is known as a _____.
   a. main wind-force-resisting system
   b. truss system
   c. secondary wind system
   d. none of the above

2. Continuous special inspection requires _____ observation of work requiring special inspection.
   a. periodic   b. full-time
   c. part-time   d. occasional

3. Sprayed fire-resistant materials are _____ materials that are spray-applied to provide fire-resistant protection of the substrates.
   a. cementitious   b. fibrous
   c. either a or b   d. none of the above

4. When working with steel construction, special inspection of the steel-fabrication process shall not be required if the fabricator does not perform any _____.
   a. welding   b. plumbing
   c. electrical work   d. none of the above

5. When working with steel construction, special inspection of the steel-fabrication process shall not be required if the fabricator does not perform any _____.
   a. welding   b. plumbing
   c. thermal cutting   d. both a and c

6. When working with steel construction, special inspection of the steel-fabrication process shall not be required if the fabricator does not perform any _____.

    a. welding

    b. plumbing

    c. heating operation of any kind

    d. both a and c

7. While work is in progress, the special inspector shall determine that the requirements for _____ and their installation and tightening standards are met.

    a. bolts

    b. nuts

    c. washers

    d. all of the above

8. Monitoring of bolt installation for pretensioning using the calibrated-wrench method or the turn-of-nut method without matchmarking shall be performed on a _____ basis.

    a. continuous

    b. periodic

    c. either a or b

    d. none of the above

9. Special inspections are not required during concrete construction utilizing isolated spread concrete footings of buildings _____ stories or less in height that are fully supported on earth or rock.

    a. 2

    b. 3

    c. 4

    d. 6

!Codealert

A designated seismic system is one where architectural, electrical, and mechanical systems and their components require design in accordance with Chapter 13 of ASCE 7 and where the component importance factor is greater than 1 in accordance with Section 13.1.3 or ASCE 7.

!**Code**alert

A main wind-force-resisting system is an assemblage of structural elements assigned to provide support and stability for the overall structure. The system generally receives wind loading from more than one surface.

10.   Special inspections are not required during concrete construction utilizing continuous concrete footings supporting walls of buildings _____ stories or less in height that are fully supported on earth or rock.

a. 2   b. 3

c. 4   d. 6

11.   Special inspections are not required during concrete construction for _____ on grade.

a. patios   b. driveways

c. sidewalks   d. all of the above

12.   Special inspections are not required for concrete construction of nonstructural concrete slabs supported directly on the ground, including prestressed slabs on grade, where the effective prestress in the concrete is less than _____ psi.

a. 150   b. 160

c. 175   d. 200

13.   Special inspections are not required for _____ masonry.

a. empirically designed   b. glass-unit

c. either a or b   d. none of the above

14.   Special inspections can be required for _____.

a. existing site soil conditions

b. fill placement

c. load-bearing requirements

d. all of the above

15. Special inspection is not required during placement of controlled fill having a total depth of _____ feet or less.

   a. 6
   b. 10
   c. 12
   d. 16

16. The cohesive/adhesive bond strength of cured, sprayed fire-resistant material applied to structural elements shall not be less than _____ pounds per square foot.

   a. 100
   b. 125
   c. 150
   d. 175

17. The acronym EIFS refers to _____.

   a. exterior insulation and finish systems
   b. external info final solutions
   c. existing introvert finish systems
   d. none of the above

18. Special inspections shall be required for proposed work that is, in the opinion of _____, unusual in its nature.

   a. the public-works official
   b. the mayor
   c. the building official
   d. none of the above

19. Qualifications for special-inspection agencies for smoke controls shall have expertise in _____.

   a. fire-protection engineering
   b. building
   c. heating systems
   d. none of the above

20. Qualifications for special-inspection agencies for smoke controls shall have expertise in _____.

   a. fire protection engineering
   b. mechanical engineering
   c. both a and b
   d. none of the above

21. Special inspections for seismic resistance require periodic inspection of _____.
    a. gluing   b. nailing
    c. both a and b   d. none of the above

22. Special inspections for seismic resistance require periodic inspection of _____.
    a. bolting   b. anchoring
    c. both a and b   d. none of the above

23. Special inspections are required during _____ welding operations of elements of the seismic-force-resisting systems for cold-formed steel framing.
    a. continuous   b. periodic
    c. overlapping   d. none of the above

24. When working with cold-formed steel framing, periodic inspection is required for _____.
    a. soldering   b. flux application
    c. screw attachment   d. none of the above

25. When working with cold-formed steel framing, periodic inspection is required for _____.
    a. bolting   b. flux application
    c. screw attachment   d. both a and c

26. When working with cold-formed steel framing, periodic inspection is required for _____.
    a. bolting   b. anchoring
    c. screw attachment   d. all of the above

27. When working with cold-formed steel framing, periodic inspection is required for _____.
    a. struts   b. anchoring
    c. screw attachment   d. all of the above

28. When working with cold-formed steel framing, periodic inspection is required for _____.
    a. bolting   b. anchoring
    c. braces   d. all of the above

> # !**Code**alert
> The special inspections for steel elements of buildings and structures shall be as required by Section 1704.3 and Table 1704.3.

29. Periodic special inspection is required during the anchorage of access floors and storage racks _____ feet or greater in height in structures assigned to seismic design categories D, E, and F.

    a. 4   b. 5

    c. 8   d. 10

30. Structures assigned to seismic design categories D, E, and F require periodic special inspections during the erection and fastening of _____.

    a. interior trim   b. exterior cladding

    c. both a and b   d. none of the above

31. Structures assigned to seismic design categories D, E, and F require periodic special inspections during the erection and fastening of _____.

    a. interior trim   b. interior nonbearing walls

    c. both a and b   d. none of the above

32. Structures assigned to seismic design categories D, E, and F do not require periodic special inspections during the erection and fastening of interior nonbearing walls weighing _____ psf or less.

    a. 15   b. 20

    c. 25   d. none of the above

33. Structural observations for seismic resistance shall be provided when the height of a structure is greater than _____ feet above the base.

    a. 25   b. 50

    c. 75   d. 100

34. A building in which a load-test procedure is being tested for design load shall, within 24 hours after removal of the test load, recover not less than _____ percent of maximum deflection.

    a. 25   b. 45

    c. 65   d. 75

35. Base metal thicker than _____ inches, where subject to through-thickness weld shrinkage strains, shall be ultrasonically tested for discontinuities behind and adjacent to such welds after joint completion.

    a. 1 1/4   b. 1 1/2

    c. 1 3/4   d. 2

TRUE-FALSE QUESTIONS

1. An established and recognized agency regularly engaged in conducting tests or furnishing inspection services, when such agency has been approved, is known as an approved agency.

True   False

2. A certificate of compliance can state that the work was done in compliance with approved construction documents.

True   False

3. Fabricated items must not be subjected to thermal cutting.

True   False

4. Main wind-force-resisting systems generally receive wind loading from no more than one surface.

True   False

!Codealert

See all of Section 1704 for numerous code amendments.

!Codealert

Sections 1705 and 1706 contain a number of changes that should be observed for code compliance.

5.　In order to obtain approval, copies of necessary test and inspection records shall be filed with the local tax adjuster.

　　True　　False

6.　Special inspections are not required for work of a minor nature.

　　True　　False

7.　Special inspections are not required for building components, unless the design involves the practice of professional engineering or architecture.

　　True　　False

8.　Occupancies in Group R-3 are required to be subject to special inspections, unless a variance is received from the local building official.

　　True　　False

9.　A special inspector shall be continuously present during all welding activities.

　　True　　False

10.　A special inspector is not required to be continuously present during the welding of stairs and railing systems.

　　True　　False

11.　The installation of high-strength bolts shall be continuously inspected during the progress of the work.

　　True　　False

12.　Monitoring of bolt installation for pretensioning is permitted to be performed on a periodic basis when using the turn-of-nut method with matchmarking techniques, the direct-tension indicator method, or the alternate design fastener method.

　　True　　False

13. Special inspections are not required for empirically designed masonry.

    True   False

14. Special inspections shall be performed during installation and testing of pile foundations.

    True   False

15. The installation of pier foundations does not require special inspections.

    True   False

16. Special inspections shall not be required for EIFS applications installed under a water-resistive barrier with a means of draining moisture to the exterior.

    True   False

17. Special inspections shall not be required for EIFS applications installed over masonry or concrete walls.

    True   False

18. Smoke-control systems shall be tested by a special inspector.

    True   False

19. Qualifications for special inspection agencies for smoke control require expertise and certification as air balancers.

    True   False

20. Continuous special inspection is required during field gluing operations of elements of the seismic-force-resisting system.

    True   False

!Codealert

Structural observations in Section 1709 have changed, so check this section to get up to speed with the current code requirements.

MULTIPLE-CHOICE ANSWER KEY

1. A	8. A	15. C	22. C	29. C
2. B	9. B	16. C	23. B	30. B
3. C	10. B	17. A	24. C	31. B
4. A	11. D	18. C	25. D	32. A
5. D	12. A	19. A	26. D	33. C
6. D	13. C	20. C	27. D	34. D
7. D	14. D	21. B	28. D	35. B

TRUE-FALSE ANSWER KEY

1. T	5. F	9. F	13. T	17. T
2. T	6. T	10. T	14. T	18. T
3. F	7. T	11. F	15. F	19. T
4. F	8. F	12. T	16. F	20. T

Chapter 18

SOILS AND FOUNDATIONS

Buildings and foundations subject to scour or water-pressure loads must be designed in accordance with this code for soils and foundations. What are those requirements? Although it may seem difficult to know everything about the code, it is required for a reason. It is easy to memorize it, but does that mean you know it? No, it does not. To know it is to apply it to construction sites on a daily basis. Take this test to see what you know and what you need to study more in regard to soils and foundations.

MULTIPLE-CHOICE QUESTIONS

1. Soil classification shall be based on observation and any necessary tests of the materials disclosed by borings, test pits, or other subsurface exploration made in appropriate locations. Additional studies shall be made as necessary to evaluate _____.

    a. slope stability
    b. slope instability
    c. soil strength
    d. both a and c

2. Soil classification shall be based on observation and any necessary tests of the materials disclosed by borings, test pits, or other subsurface exploration made in appropriate locations. Additional studies shall be made as necessary to evaluate _____.

    a. position of load-bearing capacity

    b. adequacy of load-bearing soils

    c. soil strength

    d. all of the above

3. Soil classification shall be based on observation and any necessary tests of the materials disclosed by borings, test pits, or other subsurface exploration made in appropriate locations. Additional studies shall be made as necessary to evaluate _____.

    a. effects of moisture variation on soil-bearing capacity

    b. load

    c. local rainfall rates

    d. all of the above

4. Soil classification shall be based on observation and any necessary tests of the materials disclosed by borings, test pits, or other subsurface exploration made in appropriate locations. Additional studies shall be made as necessary to evaluate _____.

    a. compressibility
    b. liquefaction
    c. expansiveness
    d. all of the above

5.   Fill that is placed, compacted, and sloped in flood-hazard areas must minimize _____.

    a. shifting   b. erosion

    c. slumping   d. all of the above

6.   Presumptive load-bearing capacity that is substantiated by data for the use of such values can result in _____ being used as load-bearing soil.

    a. mud   b. organic silt

    c. both a and b   d. none of the above

7.   Presumptive load-bearing capacity that is substantiated by data for the use of such values can result in _____ being used as load-bearing soil.

    a. organic clays   b. peat

    c. both a and b   d. none of the above

8.   Presumptive load-bearing capacity that is substantiated by data for the use of such values can result in _____ being used as load-bearing soil.

    a. unprepared fill   b. peat

    c. both a and b   d. none of the above

9.   The minimum depth of footings below the undisturbed ground surface shall be _____ inches.

    a. 8   b. 12

    c. 16   d. 24

!Codealert

Foundation design for seismic overturning refers to the requirement that the foundation be proportioned using the load combinations of Section 1605.2. The computation of the seismic overturning moment is by the equivalent lateral-force method or the modal analysis method, and the proportioning shall be in accordance with Section 12.13.4 or ASCE 7.

10. In general, buildings below slopes shall be set a sufficient distance from the slope to provide protection from _____.

   a. slope drainage

   b. sun exposure

   c. frost

   d. both a and c

11. In general, buildings below slopes shall be set a sufficient distance from the slope to provide protection from _____.

   a. slope drainage

   b. erosion

   c. frost

   d. both a and b

12. In general, buildings below slopes shall be set a sufficient distance from the slope to provide protection from _____.

   a. slope drainage

   b. erosion

   c. shallow failures

   d. all of the above

13. Where the existing slope is steeper than 100 percent, the toe of the slope shall be assumed to be at the intersection of a horizontal plane drawn from the top of the foundation and a plane drawn tangent to the slope at an angle of _____ degrees to the horizontal.

   a. 20

   b. 25

   c. 45

   d. 60

14. On graded sites, the top of any exterior foundation shall extend above the elevation of the street gutter at the point of discharge or the inlet of an approved drainage device a minimum of _____ inches plus 2 percent.

   a. 6

   b. 10

   c. 12

   d. 16

15. The minimum width of a footing shall be _____ inches.

   a. 8

   b. 12

   c. 14

   d. 16

16. Footings shall be designed for the most _____ effects due to the combinations of loads.

   a. unfavorable

   b. favorable

   c. extreme

   d. none of the above

17. Foundation walls of rough or random rubble stone shall not be less than _____ inches thick.

  a. 8
  b. 10
  c. 12
  d. 16

18. When working with concrete foundation walls, the concrete cover for reinforcement measured from the inside face of the wall shall not be less than _____ inch.

  a. 1/4
  b. 1/2
  c. 3/4
  d. 1

19. When working with concrete foundation walls, the concrete cover for reinforcement measured from the outside face of the wall shall not be less than _____ inches.

  a. 1 1/4
  b. 1 1/2
  c. 1 3/4
  d. 2

20. Concrete for foundation walls must have a specified compressive strength of not less than 2,500 psi at _____ days.

  a. 7
  b. 14
  c. 28
  d. 32

21. Masonry foundation walls require vertical reinforcement with a minimum yield strength of _____ psi.

  a. 20,000
  b. 40,000
  c. 60,000
  d. 75,000

22. Masonry foundation walls shall be built with masonry units that are installed with Type _____ mortar.

  a. M
  b. S
  c. either a or b
  d. neither a or b

23. Hollow masonry walls used for foundations require a minimum of _____ inches of solid masonry at girder supports at the top of walls.

  a. 4
  b. 6
  c. 8
  d. 12

24. Retaining walls shall be designed to ensure _____.

  a. stability
  b. sliding
  c. both a and b
  d. none of the above

25.   Retaining walls shall be designed to ensure _____.

    a. excessive foundation pressure

    b. sliding

    c. both a and b

    d. none of the above

26.   Retaining walls shall be designed to ensure _____.

    a. even temperature   b. balanced height

    c. both a and b   d. none of the above

27.   Retaining walls shall be designed to ensure _____.

    a. stability   b. sliding

    c. water uplift   d. all of the above

28.   With some exceptions walls or portions thereof that retain earth and enclose interior spaces and floors below grade shall be _____.

    a. waterproofed   b. damp-prooofed

    c. both a and b   d. none of the above

29.   For buildings and structures in flood-hazard areas, the finished ground level of an underfloor space such as a crawlspace shall be _____ the outside finished ground level.

    a. equal to   b. higher than

    c. a or b   d. a and b

30.   When installed beneath a slab, damp-proofing shall consist of not less than _____-mil polyethylene.

    a. 4   b. 6

    c. 8   d. 10

31.   When installed beneath a slab, damp-proofing shall be polyethylene installed with joints lapped not less than _____ inches.

    a. 4   b. 6

    c. 8   d. 10

> # !Codealert
> Any substantial sudden increase in rate of penetration of a timber pile shall be investigated for possible damage. If the sudden increase in rate of penetration cannot be correlated to soil strata, the pile shall be removed for inspection or rejected.

32. Damp-proofing materials for walls shall be installed on the _____ surface of the wall.

    a. interior　　　　　　　　b. exterior

    c. either a or b　　　　　　d. neither a or b

33. Damp-proofing materials for walls shall be installed to extend to the _____ of the footing to above ground level.

    a. top　　　　　　　　　　b. bottom

    c. either a or b　　　　　　d. neither a or b

34. Damp-proofing shall consist of a _____ material.

    a. soft　　　　　　　　　　b. cold

    c. bituminous　　　　　　　d. none of the above

35. Damp-proofing shall consist of a(n)_____ material.

    a. soft　　　　　　　　　　b. cold

    c. acrylic modified cement　　d. none of the above

36. Damp-proofing shall consist of a(n)_____ material.

    a. acrylic modified cement　　b. approved

    c. bituminous　　　　　　　d. any of the above

37. Holes in concrete walls created by the removal of form ties where damp-proofing is required are usually sealed with _____ materials.

    a. lead　　　　　　　　　　b. bituminous

    c. clear　　　　　　　　　　d. none of the above

38.   Parging for foundation walls shall be _____ at the footing

    a. coved   b. covered

    c. coded   d. none of the above

39.   Parging of unit masonry walls is not required when a material is approved for _____ application to the masonry.

    a. indirect   b. direct

    c. water   d. none of the above

40.   Floors required to be waterproofed shall be made of _____.

    a. marine plywood   b. concrete

    c. paving bricks   d. none of the above

41.   Walls required to be waterproofed shall be made of _____.

    a. masonry   b. concrete

    c. either a or b   d. none of the above

42.   Waterproofing on walls shall be applied from the bottom of the wall to not less than _____ inches above the maximum elevation of the ground-water table.

    a. 6   b. 12

    c. 16   d. 18

43.   A foundation drain shall be placed around the perimeter of a foundation; it must consist of gravel or crushed stone containing not more than 10 percent material that passes through a Number _____ sieve.

    a. 1   b. 2

    c. 3   d. 4

44.   When dealing with pier-and-pile foundations, the length of the pile from the first point of zero lateral deflection to the underside of the pile cap or grade beam is known as _____ length.

    a. total   b. flexural

    c. developed   d. none of the above

45. Micropiles are _____inch-diameter or less bored, grouted-in-place piles incorporating steel pipe and/or steel reinforcement.

    a. 6   b. 10

    c. 12   d. 16

46. Timber piles are _____, tapered timbers with the small end embedded into the soil.

    a. round   b. square

    c. rectangular   d. any of the above

47. Pile caps shall be made of _____.

    a. mortar   b. concrete

    c. reinforced concrete   d. masonry

48. Caisson piles are cast-in-place concrete piles extending into _____.

    a. undisturbed earth   b. water

    c. bedrock   d. any of the above

49. Augered _____ piles are constructed by depositing concrete into an uncased augered hole, either during or after the withdrawal of the auger.

    a. uncased   b. cased

    c. protected   d. none of the above

50. Piers and piles shall be braced to provide _____ stability in all directions.

    a. vertical   b. lateral

    c. fluid   d. none of the above

!Codealert

See section 1810.8 for new code regulations on micropiles.

TRUE-FALSE QUESTIONS

1.   With one exception, where the classification, strength, or compressibility of the soil is in doubt or where a load-bearing value superior to that specified in the code is claimed, a building official shall require that the necessary investigation be made.

     True   False

2.   If expansive soils are suspected, a building official shall require soil tests to determine where such soils do exist.

     True   False

3.   Waterproofing can create an exception for a soil investigation where ground water is suspected.

     True   False

4.   Pile and pier foundations shall be designed and installed on the basis of a foundation investigation and report.

     True   False

5.   When investigating rock strata for a pier foundation, borings must be taken that penetrate to a depth of not less than 8 feet below the level of the foundation.

     True   False

6.   Soil classification shall be based on observation and any necessary tests of the materials disclosed by borings, test pits, or other subsurface exploration made in appropriate locations.

     True   False

7.   A registered design professional must attend all sites during all boring and sample operations.

     True   False

8.   The soil classification and design load-bearing capacity shall be shown on the construction document.

     True   False

9.   A record of soil profile can be required in soil-classification reports.

     True   False

10. Water-table reports are not required in soil-classification reports.

     True   False

11. Excavation for any purpose shall not remove lateral support from any footing or foundation.

     True   False

12. Soil that is free of organic material is used to backfill an excavation outside a foundation.

     True   False

13. In flood-hazard areas, with some exceptions established in Section 1612.3, grading and/or fill shall not be approved.

     True   False

14. Compacted fill is not allowed to be used as a base for a footing or foundation.

     True   False

15. CLSM stands for controlled low-strength material.

     True   False

16. Presumptive load-bearing values shall apply to materials with similar physical characteristics and dispositions.

     True   False

17. For clay, sandy clay, silty clay, and clayey silt, in no case shall the lateral sliding resistance exceed one-third the dead load.

     True   False

18. The bottom surface of footings is permitted to have a slope not exceeding one unit vertical in ten units horizontal.

     True   False

19. The stepping of footings to allow for changes in elevation is not approved for single-family homes.

     True   False

20. Frost protection for footings is not required when the footings are erected on solid rock.

     True   False

21.   Where it is known that shallow subsoils are of a shifting or moving character, footings may not be installed.

       True   False

22.   Footings are prohibited from slopes 33.3-percent or greater.

       True   False

23.   Where a retaining wall is constructed for a building on an ascending slope at the toe of the slope, the height of the slope shall be measured from the top of the wall to the top of the slope.

       True   False

24.   A footing setback from a descending slope surface shall be protected from detrimental settlement.

       True   False

25.   A pool built on a slope must be capable of supporting the water in the pool without soil support.

       True   False

26.   Vibratory loads are a consideration in the construction of footings.

       True   False

27.   Balanced backfill height is the difference in height between the exterior-finish ground level and the lower of the top of the concrete footing that supports the foundation wall or the interior-finish ground level.

       True   False

28.   Under no conditions shall a foundation wall be designed to support the full hydrostatic pressure of undrained backfill.

       True   False

29.   Wood foundation plates or sills shall be bolted or strapped to the foundation being served by the wood member.

       True   False

30.   Footings placed on or within the active zone of expansive soils shall be designed to resist differential volume changes.

       True   False

MULTIPLE-CHOICE ANSWER KEY

1. D	11. D	21. C	31. B	41. C
2. D	12. D	22. C	32. B	42. B
3. A	13. C	23. A	33 A	43. D
4. D	14. C	24. C	34. C	44. B
5. D	15. B	25. C	35. C	45. C
6. C	16. A	26. D	36. D	46. A
7. C	17. D	27. D	37. B	47. C
8. C	18. C	28. C	38. A	48. C
9. B	19. B	29. C	39. B	49. A
10. A	20. C	30. B	40. B	50. B

TRUE-FALSE ANSWER KEY

1. T	7. F	13. T	19. F	25. T
2. T	8. T	14. F	20. T	26. T
3. T	9. T	15. T	21. F	27. F
4. T	10. F	16. T	22. F	28. F
5. F	11. F	17. F	23. T	29. T
6. T	12. T	18. T	24. T	30. T

Chapter 19
CONCRETE

The information contained in the code about concrete is invaluable for the construction world. There is information that ranges from definitions to pipe columns. Your job is to know which types of construction documents you need for structural concrete construction. If you have read and understand this chapter, which governs the materials, quality control, design, and construction used in concrete, you will not have an issue passing this sample test.

MULTIPLE-CHOICE QUESTIONS

1. Construction documents for structural-concrete construction shall include the specified _____ or reinforcement.

   a. grade   b. strength

   c. either a or b   d. none of the above

2. Construction documents for structural-concrete construction shall include the size and location of _____.

   a. structural elements   b. reinforcement

   c. anchors   d. all of the above

3. Construction documents for structural concrete must make provision for dimensional changes resulting from _____.

   a. creep   b. shrinkage

   c. temperature   d. all of the above

4. Construction documents for structural concrete must include the magnitude and _____ of prestressing forces.

   a. location   b. type

   c. temperature   d. all of the above

> ### !Codealert
> The format and subject mater of Sections 1902 through 1907 of this chapter are patterned after and in general conformity with the provisions for structural concrete in ACI 318.

> ## !Codealert
> See Section 1904 for changes in the code as they relate to durability requirements.

5. Construction documents for structural concrete must include anchorage length of reinforcement and location and length of _____.

   a. bolts

   b. compressive strength

   c. lap splices

   d. all of the above

6. Factors considered for curing concrete include_____.

   a. time

   b. temperature

   c. moisture conditions

   d. all of the above

7. Exposed reinforcement, inserts, and plates intended for bonding with future extensions shall be protected from _____.

   a. rain

   b. corrosion

   c. both a and b

   d. neither a or b

8. The code refers to special structures when discussing structural plain concrete. An example of a special structure is a(n)_____.

   a. arch

   b. gravity wall

   c. both a and b

   d. neither a or b

> ## !Codealert
> There have been many changes in code regulations for this chapter, so you should review the entire chapter, but pay special attention to Section 1905 on concrete quality, mixing, and placing.

> **!Code**alert
>
> Construction joints, including their location, shall comply with the provisions of ACI 318, Section 6.4.

9.   The code refers to special structures when discussing structural plain concrete. An example of a special structure is a(n) _____.

   a.   underground utility structure

   b.   shielding wall

   c.   both a and b

   d.   neither a or b

10.   The thickness of exterior basement walls and foundation walls shall be not less than _____ inches.

   a.   6 inches   b.   7 inches

   c.   7 1/2 inches   d.   8 inches

TRUE-FALSE QUESTIONS

1.   Construction documents for structural concrete shall include the specified compressive strength of concrete at the stated ages or stages of construction for which each concrete element is designed.

   True   False

> **!Code**alert
>
> The placement of reinforcement, including tolerances on depth and cover, shall comply with the provisions of ACI 318, Section 7.5. Reinforcement shall be accurately placed and adequately supported before concrete is placed.

2. Construction documents for structural concrete shall include the details and location of contraction or isolation joints specified for plain concrete.

True   False

3. Reinforcement in concrete shall not be protected from corrosion and exposure to chlorides.

True   False

4. Qualified field-testing technicians shall perform all required laboratory tests.

True   False

5. Not less than four Number 5 bars shall be provided around window and door openings.

True   False

6. A vapor retarder shall be required for all concrete slabs.

True   False

7. Shotcrete is mortar or concrete that is pneumatically projected at high velocity onto a surface.

True   False

8. Coarse aggregate used in shotcrete shall not exceed 3/4 inch.

True   False

9. Spirally tied columns require the use of shotcrete.

True   False

10. During the curing period, shotcrete shall be maintained above 40 degrees F and in moist condition.

True   False

11. Sampling specimens shall be taken from the in-place work or from test panels and shall be taken at least once during each shift and not less than one for each 75 cubic yards of shotcrete.

True   False

12. With some exceptions, the minimum thickness of reinforced gypsum concrete shall be 2 inches.

True   False

13.   Shop-fabricated concrete-filled pipe columns shall be inspected by the local plumbing inspector prior to installation.

      True   False

14.   The welding of brackets without mechanical anchorage shall be prohibited when working with concrete-filled pipe columns.

      True   False

15.   Details of column connections and splices shall be shop-fabricated by approved methods and shall be approved only after tests in accordance with the approved rules.

      True   False

!Codealert

Section 1908 contains numerous code changes.

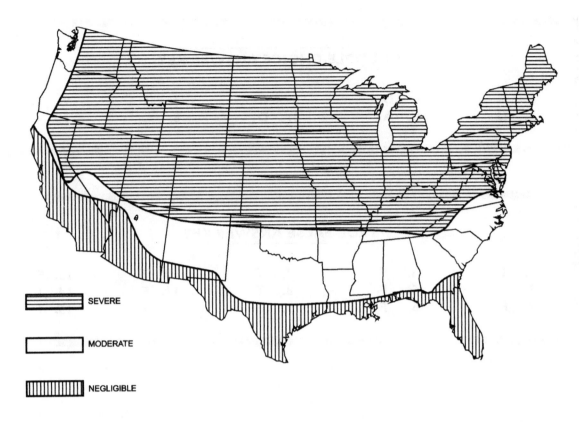

FIGURE 19.1   Weathering probability map for concrete. [a, b,c]

a.   Lines defining areas are approximate only. Local areas can be more or less severe than indicated by the region classification.
b.   A "severe" classification is where weather conditions encourage or require the use of deicing chemicals or where there is potential for a continuous presence of moisture during frequent cycles of freezing and thawing. A "moderate" classification is where weather conditions occasionally expose concrete in the presence of moisture to freezing and thawing, but where deicing chemicals are not generally used. A "negligible" classification is where weather conditions rarely expose concrete in the presence of moisture to freezing and thawing.
c.   Alaska and Hawaii are classified as severe and negligible, respectively.

MULTIPLE-CHOICE ANSWER KEY

1.	C	3.	D	5.	C	7.	B	9.	C
2.	D	4.	A	6.	D	8.	C	10.	C

TRUE-FALSE ANSWER KEY

1.	T	4.	F	7.	T	10.	T	13.	F
2.	T	5.	F	8.	T	11.	F	14.	T
3.	F	6.	F	9.	F	12.	T	15.	T

Chapter 20
ALUMINUM

Chapter 20, which covers aluminum, was brief and did not warrant a question and answer section.

Chapter 21
MASONRY

Do you know the difference between stabilized and unstabilized adobe? Are you familiar with the different dimensions? There are many topics in this chapter about masonry that you have to know in the construction world. You cannot do a professional job without this knowledge. That is why this test was designed to give you the edge that you need. Use this as a study guide for the real test to obtain a passing grade.

MULTIPLE-CHOICE QUESTIONS

1.   The construction documents shall show all of the items required by the code, such as _____.

   a. specified size   b. grade

   c.. type   d. all of the above

2.   The construction documents shall show all of the items required by the code, such as _____.

   a. location of reinforcement   b. anchors

   c. wall ties   d. all of the above

3.   The construction documents shall show all of the items required by the code, such as _____.

   a. reinforcing bars to be welded

   b. welding procedure

   c. neither a or b

   d. both a and b

4.   The construction documents shall show all of the items required by the code, such as _____.

   a. provisions for dimensional changes

   b. location of structural elements

   c. neither a or b

   d. both a and b

5.   Fireplace construction documents must describe in sufficient detail the _____ of masonry fireplaces.

   a. location   b. size

   c. neither a or b   d. both a and b

6. Construction in which the exterior load-bearing and non-load-bearing walls and partitions are of unfired clay masonry units and floors, roofs, and interior framing are wholly or partly of wood or other approved materials is known as _____.

    a. ACC masonry   b. Adobe construction

    c. balloon framing   d. none of the above

7. A masonry anchor can be made of _____.

    a. metal rod   b. wire

    c. strap   d. any of the above

8. The area of the surface of a masonry unit that is in contact with mortar in the plane of a joint is known as being _____.

    a. bedded   b. set

    c. neither a or b   d. both a and b

9. The horizontal layer of mortar on which a masonry unit is laid is known as a _____.

    a. bond beam   b. bed joint

    c. cold joint   d. none of the above

10. A horizontal grouted element within masonry in which reinforcement is embedded is known as a _____.

    a. bond beam   b. bed joint

    c. cold joint   d. none of the above

11. The adhesion between steel reinforcement and mortar or grout is known as a _____.

    a. bond beam   b. bed joint

    c. bond reinforcing   d. none of the above

12. Chimney types include _____.

    a. high-heat appliance types   b. low-heat appliance types

    c. both a and b   d. none of the above

13. Chimney types include _____.

    a. high-heat appliance types   b. low-heat appliance types

    c. masonry types   d. all of the above

> **!Code**alert
>
> Allowable stress design refers to masonry designed by the allowable stress design method, and it shall comply with the provisions of Sections 2106 and 2107.

14. Chimney types include _____.

    a. high-heat appliance types

    b. low-heat appliance types

    c. medium-heat appliance types

    d. all of the above

15. Connectors are mechanical devices for securing two or more pieces, parts, or members together, including _____.

    a. wall ties

    b. anchors

    c. either a or b

    d. both a and b

16. Connectors are mechanical devices for securing two or more pieces, parts, or members together, including _____.

    a. wall ties

    b. fasteners

    c. either a or b

    d. both a and b

17. When speaking of dimensions, the discussion could include _____.

    a. actual

    b. nominal

    c. specified

    d. all of the above

18. The opening between the top of the firebox and the smoke chamber of a fireplace is known as a _____.

    a. collar

    b. flue

    c. fireplace throat

    d. none of the above

19. A roof or floor system designed to transmit lateral forces to shear walls or other lateral-load-resisting elements is known as a _____.

    a. diaphragm   b. cover

    c. masonry unit   d. none of the above

20. A masonry unit can include _____.

    a. brick   b. tile

    c. stone   d. all of the above

21. A masonry unit can include _____.

    a. brick   b. tile

    c. glass block   d. all of the above

22. A masonry unit can include _____.

    a. brick   b. tile

    c. concrete block   d. all of the above

23. A building unit that is larger in size than a brick and composed of burned clay, shale, fire clay, or mixtures thereof is known as _____.

    a. stone   b. tile

    c. clay   d. none of the above

24. The mean daily temperature is the average daily temperature of temperature extremes predicted by a local weather bureau for the next _____ hours.

    a. 12   b. 16

    c. 24   d. none of the above

!Codealert

Glass-unit masonry shall comply with the provisions of Section 2110 or of Chapter 7 of ACI 530/ASCE5/TMS 402.

> **!Code**alert
>
> Masonry made of autoclaved aerated concrete units, manufactured without internal reinforcement and bonded together using thin- or thick-bed mortar, is AAC masonry.

25. The zone in a structural member in which the yield moment is anticipated to be exceeded under loading combinations that include earthquakes is known as _____.

    a. prestressed masonry   b. plastic hinge

    c. both a and b   d. neither a or b

26. An interior solid portion of a hollow masonry unit as placed in masonry is known as _____.

    a. a wythe   b. a web

    c. a tie   d. a shell

27. Hollow glass units shall be partially evacuated and have a minimum average glass face thickness of _____.

    a. 1/4 inch   b. 3/16 inch

    c. 1/2 inch   d. ¾ inch

28. Secondhand masonry units may be reused if they conform to the requirements of _____.

    a. new units   b. STMC C

    c. AAC   d. none of the above

29. Mill-galvanized coatings shall be applied with a minimum coating of 1/10 ounce per square foot for _____.

    a. wall ties   b. anchors

    c. inserts   d. all of the above

30. Mill-galvanized coatings for anchor bolts, steel plates, or bars not exposed to the earth, weather, or a mean relative humidity exceeding _____ percent are not required.

    a. 45
    b. 50
    c. 75
    d. 85

31. Hot-dipped galvanized coatings shall be applied with a minimum coating of 1 1/2 ounces per square foot for _____.

    a. wall ties
    b. anchors
    c. inserts
    d. all of the above

32. Hollow units shall be placed such that face shells of bed joints are fully_____.

    a. coated
    b. mortared
    c. encased
    d. covered

!Codealert

Masonry in which the tensile resistance is taken into consideration and the resistance of the reinforcing steel, if present, is neglected is known as unreinforced masonry.

33. Unless otherwise required or indicated on the construction documents, _____ units shall be placed in fully mortared bed and head joints.

    a. hollow   b. solid

    c. glass   d. none of the above

34. Masonry units shall be placed while the mortar is _____.

    a. soft   b. plastic

    c. both a and b   d. none of the above

35. Between grout pours, a horizontal construction joint shall be formed by stopping all wythes at the same elevation and with the grout stopping a minimum of _____ inches.

    a. 1 1/4   b. 1 1/2

    c. 1 3/4   d. 2

36. Wall-tie ends shall engage outer face shells of hollow units by at least_____.

    a. 1/2 inch   b. 1 inch

    c. 2 inches   d. none of the above

!Codealert

Thin-bed mortar is mortar for use in construction of AAC unit masonry with joints 0.06 inch or less.

!Codealert

See Section 2103 for current code requirements on mortar for AAC masonry.

37. Masonry directly above chases or recessed wider than _____ inches shall be supported on lintels.

    a. 10   b. 12

    c. 16   d. 24

38. The minimum length of end support with lintels shall be _____ inches.

    a. 2   b. 3

    c. 4   d. 6

39. Masonry _____ be supported on wood girders or other forms of wood construction, except as permitted in Section 2304.12

    a. shall   b. shall not

    c. may be   d. none of the above

40. Weep holes provided in the outside wythe of masonry walls shall be at a maximum spacing of _____ inches on center.

    a. 24   b. 30

    c. 33   d. 42

!Codealert

Section 2103 contains numerous code alterations, so check it out.

41.   Weep holes shall not be less than _____ in diameter.

    a.   3/16 inch   b.   4/16 inch

    c.   8/16 inch   d.   12/16 inch

42.   Temperatures of masonry units shall not be less than _____ degrees F.

    a.   32   b.   20

    c.   15   d.   10

43.   Construction requirements for temperatures between 40 degrees and 32 degrees F state that _____.

    a.   glass-unit masonry shall not be laid

    b.   glass units may be laid

    c.   mortar may not be used

    d.   none of the above

44.   Construction requirements for temperatures between 40 degrees and 32 degrees F state that water and aggregates used in mortar and grout shall not be heated above _____ degrees F.

    a.   95   b.   110

    c.   120   d.   140

45.   The design of masonry structures using allowable stress design requires that columns shall not exceed _____ feet in height.

    a.   8   b.   10

    c.   12   d.   16

!Codealert

Sections 2104 and 2105 deserve your attention for changes in code requirements.

!Codealert

Investigate code changes in Section 2107 for allowable stress design.

46. Empirical design shall not be used for exterior masonry elements that are not part of the lateral-force-resisting system and that are more than _____ feet above-ground.

    a. 16   b. 24

    c. 35   d. 40

47. The minimum thickness of rough, random, or coursed rubble stone walls shall be _____ inches.

    a. 12   b. 16

    c. 24   d. 28

48. The minimum thickness of foundation piers shall be _____ inches.

    a. 4   b. 6

    c. 8   d. 12

49. The maximum area of solid glass-block wall panels in both exterior and interior walls shall not be more than _____ square feet.

    a. 50   b. 75

    c. 100   d. 125

!Codealert

Empirical design of masonry in Section 2109 has new limitations for your review.

50. Stabilized adobe units shall be used in adobe walls for the first _____ inches above the fin-
    ished first-floor elevation.

    a. 4   b. 6

    c. 8   d. 10

51. Parapet walls constructed of adobe units shall _____.

    a. not be waterproofed   b. be waterproofed

    c. not be allowed   d. none of the above

52. The minimum thickness of exterior walls in one-story buildings shall be _____ inches.

    a. 6   b. 8

    c. 10   d. 12

53. A masonry fireplace is a fireplace constructed of _____.

    a. masonry   b. concrete

    c. either a or b   d. metal

54. Footings for masonry fireplaces and their chimneys shall be constructed of concrete or solid
    masonry at least _____ inches thick.

    a. 6   b. 12

    c. 16   d. 24

55. Footings for masonry fireplaces must extend a minimum of _____ inches beyond the face
    of the fireplace or foundation wall on all sides.

    a. 6   b. 12

    c. 16   d. 24

!**Code**alert

The minimum thickness of masonry shear walls shall be 8
inches.

56. The firebox of a masonry fireplace shall have a minimum depth of _____ inches.

    a. 12   b. 18

    c. 20   d. 24

57. Masonry over a fireplace opening shall be supported by a _____.

    a. three-layer wood header   b. lintel

    c. either a or b   d. none of the above

58. Masonry fireplaces shall be equipped with a ferrous metal damper located at least _____ inches above the top of the fireplace opening.

    a. 6   b. 8

    c. 10   d. 12

59. Smoke chamber walls shall be constructed of _____.

    a. solid masonry   b. stone

    c. concrete   d. any of the above

60. Smoke chamber walls shall be constructed of _____.

    a. solid masonry   b. hollow masonry units grouted solid

    c. concrete   d. any of the above

61. The minimum thickness of a fireplace hearth shall be _____ inches.

    a. 2   b. 4

    c. 6   d. none of the above

62. With some exceptions, any portion of a masonry fireplace located in the interior of a building shall have a clearance to combustibles of not less than _____ inches from the front faces and sides of masonry fireplaces and not less than 4 inches from the back faces of masonry fireplaces.

    a. 2   b. 3

    c. 4   d. 4 1/2

63. The firebox floor of a masonry heater shall be a minimum thickness of _____ inches of noncombustible material and be supported on a noncombustible footing and foundation.

    a. 2   b. 3

    c. 4   d. 6

> **!Code**alert
> See Section 2112 for masonry heaters to come up to speed
> with current code requirements.

64. Combustible materials shall not be placed within _____ inches of the outside surface of a masonry heater.

    a. 10   b. 16

    c. 24   d. 36

65. When two or more flues are located in the same chimney, masonry wythes shall be built between adjacent flue linings. The masonry wythes shall be at least _____ inches thick and bonded into the walls of the chimney.

    a. 2   b. 4

    c. 5   c. 6

TRUE-FALSE QUESTIONS

1. Construction documents are required to include the size and location of structural elements.

   True   False

2. AAC masonry is masonry made of autoclaved aerated concrete units, manufactured without internal reinforcement and bonded together using thin- or thick-bed mortar.

   True   False

3. Architectural terracotta can be plain or ornamental hard-burned modified clay units that are smaller in size than brick, with a glazed or unglazed ceramic finish.

   True   False

4. A masonry unit made of sand or clime is a calcium-silicate brick.

   True   False

5. A project part of a masonry wall that is built integrally to provide vertical stability is a buttress.

   True   False

6. A void space having a gross cross-sectional area greater than 1 square inch is known as a cell.

   True   False

7. A building stone manufactured from Portland cement precast concrete and used as a trim, veneer, or facing on or in buildings or structures is known as cast stone.

   True   False

8. A device that is primarily a vertical enclosure containing one or more passageways for conveying flue gases to the outside atmosphere is known as a chimney.

   True   False

9. A cleanout is an opening to the bottom of a grout space of sufficient size and spacing to allow the removal of debris.

   True   False

10. A vertical longitudinal joint between wythes of masonry or between masonry and backup construction that is permitted to be filled with mortar or grout is known as a collar joint.

    True   False

11. An isolated vertical member whose horizontal dimension measured at right angles to its thickness does not exceed three times its thickness and whose height is at least four times its thickness is a masonry column.

    True   False

!Codealert

See Section 2113.16.2 for net cross-sectional area of square and rectangular flue size requirements.

12. Multiwythe masonry members acting with composite action are known as composite masonry.

    True   False

13. The distance between the surface of a reinforcing bar and the edge of a member is known as cover.

    True   False

14. Masonry composed of glass units bonded by mortar is known as glass-unit masonry.

    True   False

15. A masonry unit that connects one or more adjacent wythes of masonry is known as a header.

    True   False

16. A building unit or block larger in size than 10 inches x 4 inches x 4 inches that is made of cement and suitable aggregates is known as concrete.

    True   False

17. Solid masonry refers to solid masonry units laid contiguously with the joints between the units filled with mortar.

    True   False

18. Mortar is a plastic mixture of approved cementitious materials, fine aggregates, and water used to bond masonry or other structural units.

    True   False

19. Masonry composed of roughly shaped stone is known as rubble masonry.

    True   False

20. A running bond is the placement of masonry units such that head joints in successive courses are horizontally offset at least one-quarter the unit length.

    True   False

21. A shell is the inner portion of a hollow masonry unit.

    True   False

22. A wall tie is a connector that connects wythes of masonry walls together.

    True   False

23. A lateral tie is a loop of reinforcing bar or wire enclosing longitudinal reinforcement.

True   False

24. Each continuous vertical section of a wall one masonry unit in thickness is known as a wythe.

True   False

25. To obtain design strength, one must multiply the nominal strength by a strength-reduction factor.

True   False

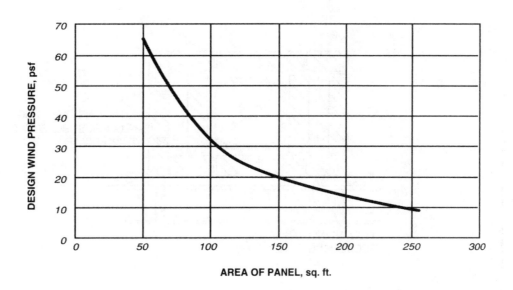

For SI: 1 square foot = 0.0929 m², 1 pound per square foot = 47.9 N/m².

FIGURE 21.1   Glass masonry design wind load resistance.

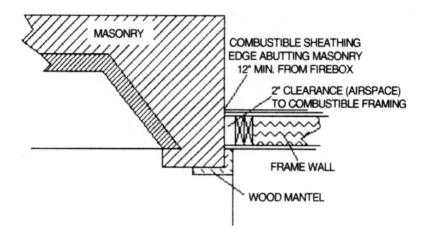

For SI: 1 inch = 25.4 mm

FIGURE 21.2   Illustration of exception to fireplace clearance provision.

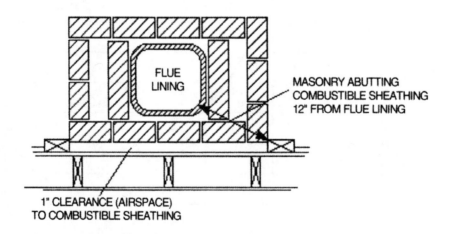

For SI: 1 inch = 25.4 mm.

FIGURE 21.3   Illustration of exception three chimney clearance provision.

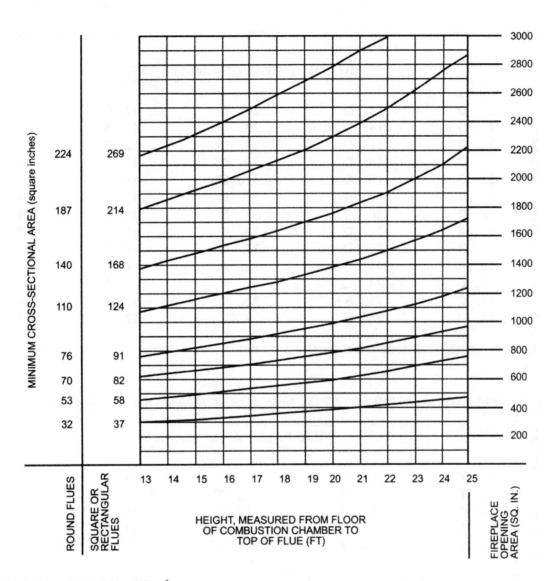

For SI: 1 inch = 25.4 mm, 1 square inch = 645 mm².

FIGURE 21.4 Flue sizes for masonry chimneys.

MULTIPLE-CHOICE ANSWER KEY

1. D	14. D	27. B	40. C	53. C
2. D	15. D	28. A	41. A	54. B
3. D	16. D	29. D	42. B	55. A
4. D	17. D	30. C	43. A	56. C
5. D	18. C	31. D	44. D	57. B
6. B	19. A	32. B	45. C	58. B
7. D	20. D	33. B	46. C	59. D
8. A	21. D	34. C	47. B	60. D
9. B	22. D	35. B	48. C	61. B
10. A	23. C	36. A	49. C	62. A
11. C	24. C	37. B	50. A	63. C
12. C	25. B	38. C	51. B	64. D
13. D	26. B	39. B	52. C	65. B

TRUE-FALSE ANSWER KEY

1. T	6. F	11. T	16. F	21. F
2. T	7. T	12. T	17. T	22. T
3. F	8. T	13. T	18. T	23. T
4. T	9. T	14. T	19. T	24. T
5. F	10. F	15. F	20. T	25. T

Chapter 22
STEEL

It is easy to be overwhelmed when preparing for a big test. You can, however, relieve some of that pressure by studying, studying, and more studying. When you are ready, take this sample test to find out how well you are doing. You may surprise yourself by how much you have learned and see that, yes, you can, pass this test with ease. This test covers the provisions that govern the design, fabrication, and erection of steel used in buildings or structures. Not only do you need to know about the provisions of the code regarding steel but also the relationship to other codes related to construction. This test is designed with you, the student, in mind.

MULTIPLE-CHOICE QUESTIONS

1. Cold-formed steel construction is made up entirely or in part of steel structural members that have been cold-formed to shape from _____ or strip steel.

    a. sheet   b. dipped

    c. both a and b   d. none of the above

2. Cold-formed steel construction is made up entirely or in part of steel structural members that may include _____.

    a. studs   b. floor joists

    c. both a and b   d. none of the above

3. Cold-formed steel construction is made up entirely or in part of steel structural members that may include _____.

    a. roof decks   b. floor panels

    c. both a and b   d. none of the above

4. Cold-formed steel construction is made up entirely or in part of steel structural members that may include _____.

    a. floor panels   b. wall panels

    c. both a and b   d. none of the above

5. A steel joist is any steel structural member of a building or structure that is made of _____.

    a. hot-rolled steel   b. rolled steel

    c. either a or b   d. none of the above

6. The design and installation of cold-formed-steel box headers used in single-span conditions for load-carrying purposes shall be in accordance with AISI-Header, subject to the limitations therein. Other elements that are required to meet the same regulations include _____.

   a. back-to-back headers
   b. single L-headers
   c. double L-headers
   d. all of the above

7. The AISI-Truss regulations apply to _____.

   a. truss design
   b. quality assurance
   c. installation of trusses
   d. all of the above

8. ASCE 3 controls the design and construction of _____.

   a. composite slabs of concrete
   b. steel deck
   c. both a and b
   d. none of the above

9. The design and installation of cold-formed-steel studs for structural and nonstructural walls shall be in accordance with _____.

   a. AISI-WSD
   b. SJI K-1.1
   c. SJI JG-1.1
   d. none of the above

10. When working with steel joists, special loads include _____.

    a. concentrated loads
    b. connection forces
    c. axial loads
    d. all of the above

TRUE-FALSE QUESTIONS

1. Riveted or welded bars can be used to make a steel joist.
   True   False

2. A structural-steel member must be made with cold-formed steel.
   True   False

3. Structural-steel members must not be made with steel joist members.
   True   False

> **!Code**alert
>
> See Section 2206 for changes in code regarding steel joists.

4.   Anchor rods shall be set accurately to the pattern and dimensions called for on approved construction plans.

   True   False

5.   The protrusion of threaded ends through the connected material shall be sufficient to fully engage the threads of the nuts but shall not be greater than the length of the threads on the bolts.

   True   False

6.   Structural steel shall not be painted.

   True   False

7.   Enameling is one method of protecting steel from corrosion.

   True   False

8.   The design, fabrication, and erection of structural steel for buildings and structures shall be in accordance with AISC 360.

   True   False

9.   Prescriptive framing applies to detached one- and two-family dwellings and townhouses up to two stories in height.

   True   False

10.   The design of light-framed cold-formed steel walls and diaphragms to resist wind and seismic loads shall be in accordance with AISI-Lateral.

   True   False

MULTIPLE-CHOICE ANSWER KEY

1. A	3. C	5. C	7. D	9. A
2. C	4. C	6. D	8. C	10. D

TRUE-FALSE ANSWER KEY

1. T	3. T	5. T	7. T	9. T
2. F	4. T	6. F	8. T	10. T

Chapter 23
WOOD

Wood is one of the longest chapters in the code book. Why? Because there are many aspects of this material that cannot be taken for granted. For further information I have included references to the standards for various wood and wood-based products. I cannot stress enough the importance of understanding all the parts and sections of the code; with such understanding they can be applied in such a way that they will not harm you, your building, or the general public. Take this sample test to see where you stand in knowing the code.

MULTIPLE-CHOICE QUESTIONS

1. An approved third-party organization that is independent of the grading and inspection agencies and the lumber mills and that initially accredits and subsequently monitors on a continuing basis the competence and performance of a grading or inspection agency in carrying out a specific task is known as a _____.

    a. code-enforcement officer   b. accreditation body

    c. accredited reviewer   d. general contractor

2. A framed stud wall extending from the top of a foundation to the underside of floor framing for the lowest occupied floor level is known as a _____.

    a. subwall   b. sill

    c. cripple wall   d. none of the above

3. A fibrous, homogeneous panel made from lignocellulosic fibers and having a density of less than 31 pounds per cubic foot but more than 10 pounds per cubic foot is known as _____.

    a. drywall   b. gypsum board

    c. fiberboard   d. either a or b

4. The classification of lumber in regard to strength and utility in accordance with the American Softwood Lumber Standard is known as _____.

    a. lumber type   b. lumber grade

    c. neither a or b   d. either a or b

5. A fibrous-felted, homogeneous panel made from lignocellulosic fibers consolidated under heat and pressure in a hot press to a density of not less than 31 pounds per cubic foot is known as _____.

    a. drywall   b. gypsum board

    c. fiberboard   d. hardboard

6. A generic term for a panel primarily composed of cellulosic materials, usually wood, generally in the form of discrete pieces or particles as distinguished from fibers is _____.

    a. particleboard   b. gypsum board
    c. fiberboard   d. hardboard

7. A wall designed to resist lateral forces parallel to the plane of a wall is known as _____.

    a. cripple wall   b. shear wall
    c. half wall   d. none of the above

8. A structural member manufactured using wood elements bonded together with exterior adhesives is known as _____.

    a. structural composite lumber

    b. I-beam

    c. subdiaphragm

    d. treated lumber

9. Treated wood is wood that is _____ compounds that reduce its susceptibility to flame spread or to deterioration caused by fungi, insects, or marine borers.

    a. soaked in

    b. sprayed with

    c. impregnated under pressure with

    d. colored with

!Codealert

The design of structural elements or systems constructed partially or wholly of wood or wood-based products must be in accordance with one of the following methods:

- Allowable stress design in accordance with Sections 2304, 2305 and 2306

- Load and resistance factor design in accordance with Sections 2304, 2305 and 2307

- Conventional light-frame construction in accordance with Sections 2304 and 2308.

!Codealert

See Section 2302 for updates of definitions regarding wood.

10. When jointing fiberboard, the joints must be tight-fitting. The approved materials for this type of work include _____.

    a. square joints   b. tongue-and-groove joints

    c. U-shaped joints   d. all of the above

11. Fiberboard wall insulation applied on the exterior of foundation walls shall be protected below ground level with _____.

    a. rigid foam insulation   b. a bituminous coating

    c. both a and b   d. none of the above

12. Treated wood is to be stamped or labeled with the following information _____.

    a. type of preservative used

    b. color code of the wood

    c. end use for which the product is treated

    d. both a and c

13. Treated wood is to be stamped or labeled with the following information _____.

    a. type of preservative used

    b. identification of treating manufacturer

    c. end use for which the product is treated

    d. all of the above

!Codealert

Section 2303.1 has added some features that you should be sure to check out.

> **!Code**alert
>
> Stress grading of structural log members of nonrectangular shape, as typically used in log buildings, must be used in accordance with ASTM D 3957, and round timber poles and piles must comply with ASTM D 3200 and ASTM D 25.

14. Treated wood is to be stamped or labeled with the following information _____.

    a. AWPA standard to which the product was treated

    b. age of the wood

    c. end use for which the product is treated

    d. both a and c

15. Moisture content for treated wood can be a concern. Where preservative-treated wood is used in an enclosed location and where drying in service cannot readily occur, the wood shall be kept at a moisture content of _____ percent or less before being covered with insulation, interior-wall finishes, floor coverings, or other materials.

    a. 7   b. 15

    c. 19   d. 23

16. Fire-retardant-treated lumber and wood structural panels are required to be labeled, and the label must contain the following information _____.

    a. the species of wood treated

    b. the flame-spread index

    c. the smoke-developed index

    d. all of the above

> **!Code**alert
>
> Section 2303.4 has some additions about trusses that would be worthy of your attention.

!Codealert

Check out Section 2304.8 for numerous changes in the lumber decking code.

17. Fire-retardant-treated lumber and wood structural panels are required to be labeled, and the label must contain the following information _____.

a. the species of wood treated

b. the flame-spread index

c. the method of drying after treatment

d. all of the above

18. When fire-retardant-treated wood is exposed to weather or damp or wet conditions, it shall be identified as _____ to indicate that there is no increase in the listed flame-spread index.

a. exterior   b. interior

c. acceptable   d. none of the above

19. Fire-retardant-treated wood structural panels must be dried to a maximum moisture content of _____ percent before they are installed.

a. 10   b. 15

c. 19   d. 25

!Codealert

Table 2304.9.1 on fastening schedule has many changes that deserve your attention.

!**Code**alert
Wood used above ground in locations specified in Sections 2304.11.2.1 through 2304.11.2.7, 2304.11.3, and 2304.11.5 must be naturally durable wood or preservative-treated wood using water-borne preservatives, in accordance with AWPA U1 for above-ground use.

20. Truss design drawings shall include, at a minimum, the following information _____.

   a. location of joints   b. required bearing widths

   c. both a and b   d. none of the above

21. Truss design drawings shall include, at a minimum, the following information _____.

   a. location of joints   b. required bearing widths

   c. bottom-chord live load   d. all of the above

22. Truss design drawings shall include, at a minimum, the following information _____.

   a. bottom-chord live load   b. required bearing widths

   c. bottom-chord dead load   d. all of the above

!**Code**alert
Section 2304.11.4 contains a variety of code changes of which you need to be aware.

!Codealert

Table 2305.2.2 (1) and (2) showing values of Gt for calculating deflection of wood structural panel walls and diaphragms includes new information that requires your attention.

23. Truss design drawings shall include, at a minimum, the following information _____.

    a. lumber size, species, and grade

    b. installer address

    c. crane size needed for installation

    d. all of the above

24. A truss submittal package is required to contain _____.

    a. each individual truss design drawing

    b. an index sheet

    c. both a and b

    d. none of the above

25. All transfer of loads and anchorage of each truss to the supporting structure is the responsibility of _____.

    a. the general contractor

    b. the master carpenter

    c. the registered design professional

    d. all of the above

!Codealert

Many changes have taken place in regards to equations for configuring the design of wood shear walls and must be looked at in Section 2305.3.

> **!Code**alert
>
> There are code changes in Section 2307 for load- and resist-
> ance-factor design to be aware of.

26. Without the written concurrence and approval of a registered design professional, truss members and components shall not be _____.

    a. cut   b. notched

    c. drilled   d. all of the above

27. Consideration shall be given in design to the possible effect of cross-grain dimensional changes considered vertically, which may occur in lumber fabricated in a green condition is known as _____.

    a. shrinkage   b. failure

    c. cold storage   d. kiln condition

28. Headers, double joists, trusses, or other approved assemblies that are of adequate size to transfer loads to vertical members shall be provided over_____ in load-bearing walls and partitions.

    a. windows   b. doors

    c. both a and b   d. none of the above

29. When wood boards are used as sheathing, the minimum thickness of the boards is _____.

    a. 1/2 inch   b. 3/4 inch

    c. 5/8 inch   d. none of the above

> **!Code**alert
>
> In Table 2306.4.1 there are numerous code changes for allow-
> able shear for wood structural-panel shear walls to take note of.

> **!Code**alert
> In Section 2308.1.1 the term "portions" means parts of build-
> ings containing volume and area, such as a room or a series of
> rooms.

30. Fiberboard that is used as sheathing must have a minimum thickness of _____.

   a. 1/2 inch   b. 3/4 inch

   c. 5/8 inch   d. none of the above

31. Gypsum wall board used as sheathing requires a minimum thickness of _____.

   a. 1/2 inch   b. 3/4 inch

   c. 5/8 inch   d. none of the above

32. Reinforced cement mortar that is used as sheathing is required to have a minimum thickness
     of _____.

   a. 1/2 inch   b. 3/4 inch

   c. 5/8 inch   d. 1 inch

33. Maximum wall-stud spacing, on center, when wood boards are used as sheathing is
     _____.

   a. 12 inches   b. 16 inches

   c. 24 inches   d. none of the above

34. Maximum wall-stud spacing, on center, when fiberboard is used as sheathing is _____.

   a. 12 inches   b. 16 inches

   c. 24 inches   d. none of the above

35. Maximum wall-stud spacing, on center, when gypsum sheathing is used is _____.

   a. 12 inches   b. 16 inches

   c. 24 inches   d. none of the above

36. Maximum wall-stud spacing, on center, when gypsum wallboard is used as sheathing is
_____.

    a. 12 inches   b. 16 inches

    c. 24 inches   d. none of the above

37. Maximum wall-stud spacing, on center, when reinforced cement mortar is used as sheathing
is _____.

    a. 12 inches   b. 16 inches

    c. 24 inches   d. none of the above

38. If deck supports are 48 inches or less, side nails shall be spaced not more than _____
inches on center.

    a. 16   b. 24

    c. 30   d. 36

39. If deck supports are spaced more than 48 inches on center, side nails shall be spaced not
more than _____ inches on center.

    a. 16   b. 18

    c. 24   d. 30

40. Two-inch sawn tongue-and-groove decking shall have a maximum moisture content of
_____ percent.

    a. 10   b. 15

    c. 20   d. 25

41. Two-inch sawn tongue-and-groove decking requires the use of _____ nails.

    a. 8d common   b. 8d galvanized

    c. 16d common   d. any of the above

!Codealert

See Section 2308.4 for changes in design of elements exceeding limitations of conventional construction and structural elements or systems not described herein.

42. Two-inch sawn tongue-and-groove decking, when installed with a controlled random pattern, shall have a minimum distance of _____ inches between end joints in adjacent courses.

    a. 12   b. 16

    c. 24   d. 32

43. Which location is allowed for the installation of a continuous header to a stud?

    a. toenail   b. face nail

    c. end nail   d. any of the above

44. Which location is allowed for the installation of double studs?

    a. toenail   b. face nail

    c. end nail   d. any of the above

45. Which location is allowed for the installation of a rim joist to a top plate?

    a. toenail   b. face nail

    c. end nail   d. any of the above

46. Which location is allowed for the installation of a collar tie to a rafter?

    a. toenail   b. face nail

    c. end nail   d. any of the above

47. Which location is allowed for the installation of ceiling joists to parallel rafters?

    a. toenail   b. face nail

    c. end nail   d. any of the above

48. Wood framing members, including wood sheathing, that rest on exterior foundation walls and are less than _____ inches from exposed earth shall be of naturally durable or preservative-treated wood for protection against termites.

    a. 6   b. 8

    c. 12   d. 16

49. Sleepers and sills on a concrete or masonry slab that is in direct contact with earth shall be of _____ wood.

    a. naturally durable   b. preservative-treated

    c. either a or b   d   none of the above

50. Posts or columns on a concrete or masonry slab that is in direct contact with earth shall be of _____ wood.

    a. naturally durable   b. preservative-treated

    c. either a or b   d. none of the above

51. Which of the following can be used as sheathing?

    a. OSB   b. 3-ply plywood

    c. 5-ply plywood   d. any of the above

52. Trimmer and header joists shall be doubled or of lumber of equivalent cross section where the span of the header exceeds _____ feet.

    a. 2   b. 3

    c. 4   d. 6

53. The spacing of 2 x 4 wall studs used for nonbearing walls in light construction shall not exceed _____ inches on center.

    a. 16   b. 24

    c. 30   d. 36

54. The spacing of 2 x 3 wall studs used for nonbearing walls in light construction shall not exceed _____ inches on center.

    a. 16   b. 24

    c. 30   d. 36

> **!Code**alert
>
> Prefabricated wood I-joists, structural glued-laminated timber, and structural composite lumber must not be notched or drilled except where permitted by the manufacturer's recommendations or where the effects of such alterations are specifically considered in the design of the member by a registered design professional.

55. The spacing of 2 x 6 wall studs used for nonbearing walls in light construction shall not exceed _____ inches on center.

    a. 16   b. 24

    c. 30   d. 36

56. Studs shall have full bearing on a plate or sill not less than _____ inches in thickness and with a width not less than that of the wall studs.

    a. 1   b. 2

    c. 3   d. 4

57. Foundation cripple walls shall be framed with studs not less in size than the studding above, with a minimum length of _____ inches, or shall be framed of solid blocking.

    a. 12   b. 14

    c. 16   d. 24

58. Spacing of edge nailing for required wall bracing shall not exceed _____ inches on center along the foundation and top plates of a cripple wall.

    a. 6   b. 8

    c. 12   d. 16

59. In exterior walls and bearing partitions, any wood stud is permitted to be cut or notched to a depth not exceeding _____ percent of the stud width.

    a. 10   b. 15

    c. 25   d. 33

60.   In no case shall the edge of a bored hole be nearer than ___ inch to the edge of the stud.

   a. 1/4   b. 1/2

   c. 5/8   d. none of the above

TRUE-FALSE QUESTIONS

1.   A collector is a horizontal diaphragm element parallel and in line with the applied force that collects and transfers diaphragm shear forces to the vertical elements of the lateral-force-resisting system and/or distributes forces within the diaphragm.

   True   False

2.   Conventional light-frame wood construction is a type of construction whose primary structural elements are formed by a system of repetitive wood-framing members.

   True   False

3.   A collector and a drag strut are the same.

   True   False

4.   An unblocked diaphragm is one that has edge nailing at supporting members only.

   True   False

5.   A nonstructural element that is composed of built-up lumber is known as a glued built-up member.

   True   False

6.   Edge nailing is a special nailing pattern at the edges of each panel within the assembly of a diaphragm or shear wall.

   True   False

7.   A prefabricated-wood I-joist is a nonstructural member that can only be used to sister up to existing joists.

   True   False

8.   A wood-structural-panel sheathed wall with openings that have not been specifically designed and detailed for force transfer is known as a perforated shear wall.

   True   False

9. LVL stands for laminated-veneer lumber.

    True   False

10. A device used to resist uplift of the chords of shear walls is a tie-down.

    True   False

11. Plywood is a wood structural panel comprised of piles of wood veneer arranged in cross-aligned layers.

    True   False

12. A subdiaphragm is a portion of a larger wood diaphragm designed to anchor and transfer local forces to secondary diaphragm struts.

    True   False

13. When jointing fiberboard, you must ensure a tight-fitting assembly.

    True   False

14. Fiberboard that is used as roof insulation in all types of construction must be provided with an approved roof covering.

    True   False

15. The quality mark for the Lumber Standards Treated Wood Program must be on a stamp or label affixed to the preservative-treated wood.

    True   False

16. Truss manufacturers shall provide a truss-placement diagram that identifies the proposed location for each individually designated truss.

    True   False

!Codealert

Changes in Table 2308.12.4 regarding wall bracing in seismic-design categories have seen some changes; it would be worth your time to check them out.

17. When truss-placement diagrams are prepared under the direct supervision of a registered design professional, the design does not have to be sealed.

    True   False

18. Computations to determine the required sizes of members shall be based on nominal sizes.

    True   False

19. Bottom plates shall have studs that have full bearing on a 2-inch thick or larger plate or sill with a width that is at least equal to the width of the studs.

    True   False

20. When working with lumber decking, a simple span pattern requires that all pieces be supported by two supports.

    True   False

21. When dealing with lumber decking, a two-span continuous pattern is required to be supported by four supports.

    True   False

22. When dealing with lumber decking on a two-span continuous pattern, all end joints shall occur in line on every other support.

    True   False

23. When dealing with lumber decking, courses in end spans shall not be alternating simple-span and two-span continuous.

    True   False

24. When floor joists are spaced 24 inches on center, the minimum net thickness is 3/4 inch for floor sheathing.

    True   False

25. The length of nails connecting laminations shall not be less than two and one-half times the net thickness of each lamination.

    True   False

26. Sheathing nails shall not be driven in so that their head or crown is flush with the surface of the sheathing.

    True   False

27. Where wall-framing members are not continuous from foundation sill to roof, the members shall be secured to ensure a continuous load path.

    True   False

28. Heavy timber construction is not permitted by the International Building Code.

    True   False

29. The location for fastening when installing a top plate to a stud is a toenail.

    True   False

30. Double top plates require typical face nailing.

    True   False

31. Where supported by a wall, roof decks shall be anchored to resist uplift forces.

    True   False

32. Untreated wood is permitted when such wood is continuously and entirely below the ground-water level or submerged in fresh water.

    True   False

33. In geographical areas where termite damage is known to be very heavy, wood floor framing must be protected from termite infestation.

    True   False

34. General design requirements for lateral-force-resisting systems require framing members to have a minimal nominal width of at least 2 inches.

    True   False

35. Wood members shall not be permitted to resist horizontal seismic forces from nonstructural concrete, masonry veneer, or concrete floors.

    True   False

36. Calculations for diaphragm deflection shall account for the usual bending and shear components.

    True   False

37. Nail deformation can contribute to deflection.

    True   False

38. Fiberboard may be used to resist seismic forces in structures in seismic design categories D, E, or F.

True   False

39. Light-frame construction requires a bearing wall with floor-to-floor heights not to exceed 10 feet, plus a height of not more than 16 inches for floor framing.

True   False

40. Braced wall lines shall be supported by continuous foundations.

True   False

41. Headers are required over each opening in exterior-bearing walls.

True   False

42. Openings in nonbearing partitions are not permitted to be framed with single studs and headers.

True   False

43. Stud partitions that will contain plumbing pipes must be framed and the joists underneath so spaced as to give proper clearance for the piping.

True   False

44. Roof assemblies shall have rafter and truss ties to the wall below.

True   False

45. Rafters shall be framed directly opposite each other at the ridge.

True   False

!Codealert

Steel-plate washers must be a minimum of 0.229 by 3 inches by 3 inches in size. The hole in the plate washer is allowed to be diagonally slotted with a width of up to 3/16 inch larger than the bolt diameter and a slot length not to exceed 1 3/4 inches, provided a standard cut washer is placed between the plate washer and the nut.

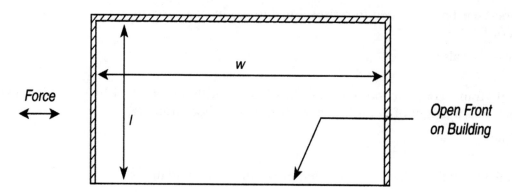

FIGURE 23.1   Diaphragm length and width for plan view of open-front building.

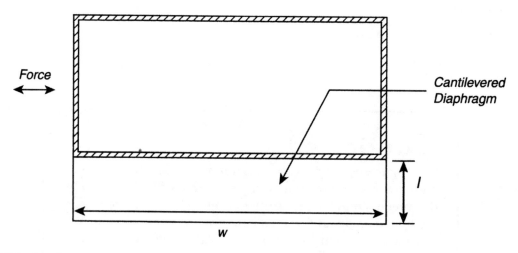

FIGURE 23.2   Diaphragm length and width for plan view of cantilevered diaphragm.

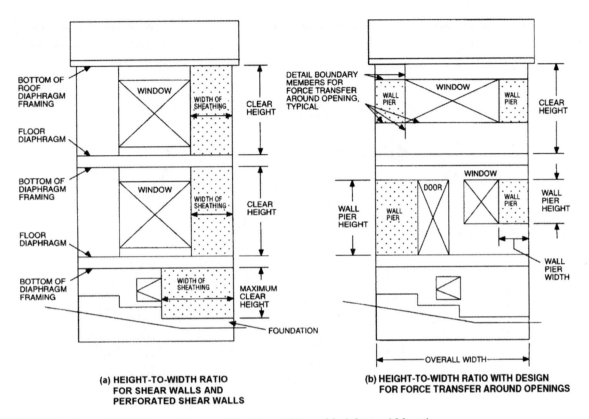

(a) HEIGHT-TO-WIDTH RATIO
FOR SHEAR WALLS AND
PERFORATED SHEAR WALLS

(b) HEIGHT-TO-WIDTH RATIO WITH DESIGN
FOR FORCE TRANSFER AROUND OPENINGS

FIGURE 23.3 General definition of shear wall height, width, and height-to-width ratio.

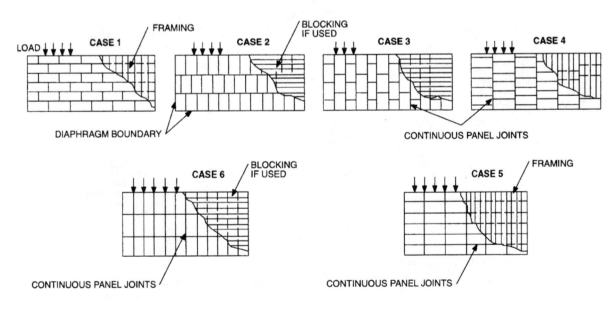

FIGURE 23.4 Allowable shear examples as they may appear in your code book.

SEISMIC DESIGN CATEGORY	MAXIMUM WALL SPACING (feet)	REQUIRED BRACING LENGTH, b
A, B and C	35'-0"	Table 2308.9.3(1) and Section 2308.9.3
D and E	25'-0"	Table 2308.12.4

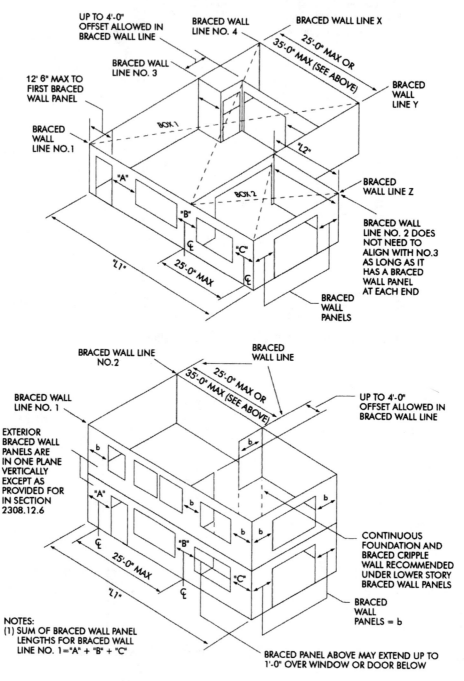

For SI: 1 foot = 304.8 mm.

FIGURE 23.5   Basic components of the lateral bracing system.

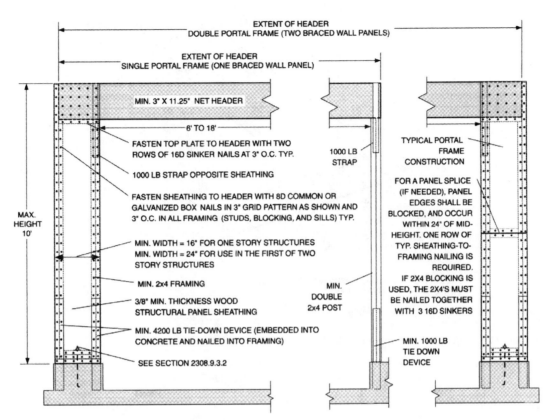

EXTENT OF HEADER
DOUBLE PORTAL FRAME (TWO BRACED WALL PANELS)

EXTENT OF HEADER
SINGLE PORTAL FRAME (ONE BRACED WALL PANEL)

MIN. 3" X 11.25" NET HEADER

6' TO 18'

FASTEN TOP PLATE TO HEADER WITH TWO
ROWS OF 16D SINKER NAILS AT 3" O.C. TYP.

1000 LB STRAP OPPOSITE SHEATHING

FASTEN SHEATHING TO HEADER WITH 8D COMMON OR
GALVANIZED BOX NAILS IN 3" GRID PATTERN AS SHOWN AND
3" O.C. IN ALL FRAMING (STUDS, BLOCKING, AND SILLS) TYP.

MIN. WIDTH = 16" FOR ONE STORY STRUCTURES
MIN. WIDTH = 24" FOR USE IN THE FIRST OF TWO
STORY STRUCTURES

MIN. 2x4 FRAMING

3/8" MIN. THICKNESS WOOD
STRUCTURAL PANEL SHEATHING

MIN. 4200 LB TIE-DOWN DEVICE (EMBEDDED INTO
CONCRETE AND NAILED INTO FRAMING)

SEE SECTION 2308.9.3.2

MAX.
HEIGHT
10'

1000 LB
STRAP

MIN.
DOUBLE
2x4 POST

MIN. 1000 LB
TIE DOWN
DEVICE

TYPICAL PORTAL
FRAME
CONSTRUCTION

FOR A PANEL SPLICE
(IF NEEDED), PANEL
EDGES SHALL BE
BLOCKED, AND OCCUR
WITHIN 24" OF MID-
HEIGHT. ONE ROW OF
TYP. SHEATHING-TO-
FRAMING NAILING IS
REQUIRED.
IF 2X4 BLOCKING IS
USED, THE 2X4'S MUST
BE NAILED TOGETHER
WITH 3 16D SINKERS

For SI: 1 foot = 304.8 mm; 1 inch = 25.4 mm; 1 pound = 4.448 N.

FIGURE 23.6   Alternate braced wall panel adjacent to a door or window opening.

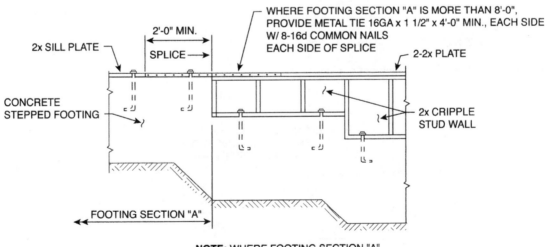

2x SILL PLATE

CONCRETE
STEPPED FOOTING

2'-0" MIN.

SPLICE

WHERE FOOTING SECTION "A" IS MORE THAN 8'-0",
PROVIDE METAL TIE 16GA x 1 1/2" x 4'-0" MIN., EACH SIDE
W/ 8-16d COMMON NAILS
EACH SIDE OF SPLICE

2-2x PLATE

2x CRIPPLE
STUD WALL

FOOTING SECTION "A"

NOTE: WHERE FOOTING SECTION "A"
IS LESS THAN 8'-0" LONG IN A
25'-0" TOTAL LENGTH WALL, PROVIDE
BRACING AT CRIPPLE STUD WALL

For SI: 1 inch = 25.4 mm, 1 foot = 304.8 mm.

FIGURE 23.7   Stepped footing connection details.

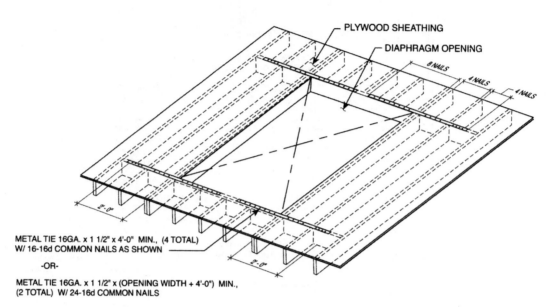

METAL TIE 16GA. x 1 1/2" x 4'-0" MIN., (4 TOTAL)
W/ 16-16d COMMON NAILS AS SHOWN

-OR-

METAL TIE 16GA. x 1 1/2" x (OPENING WIDTH + 4'-0") MIN.,
(2 TOTAL) W/ 24-16d COMMON NAILS

For SI: 1 inch = 25.4 mm, 1 foot = 304.8 mm.

FIGURE 23.8   Openings in horizontal diaphragms.

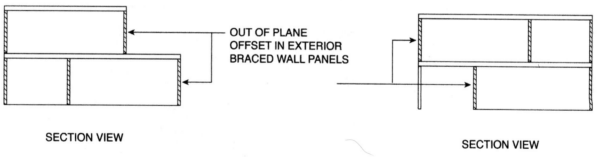

FIGURE 23.9   Braced wall panels out of plane.

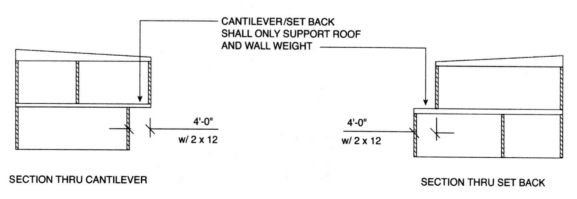

For SI: 1 foot = 304.8 mm.

FIGURE 23.10   Braced wall panels supported by cantilever or set back.

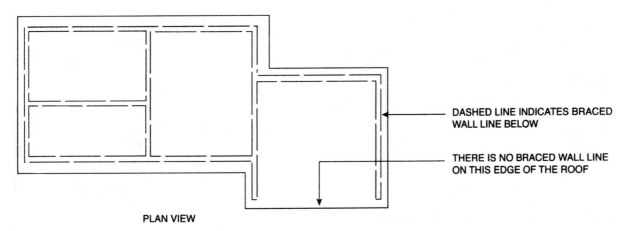

FIGURE 23.11 Floor or roof not supported on all edges.

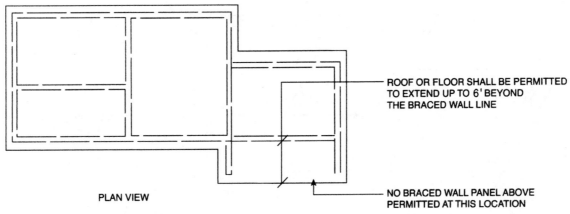

For SI: 1 foot = 304.8 mm.

FIGURE 23.12 Roof or floor extension beyond braced wall line.

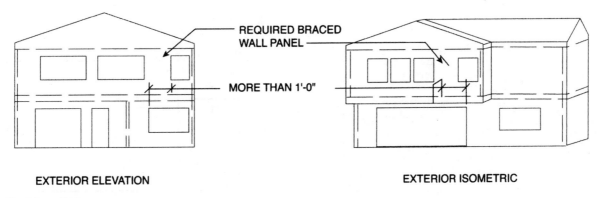

For SI: 1 foot = 304.8 mm.

FIGURE 23.13 Braced wall panel extension over opening.

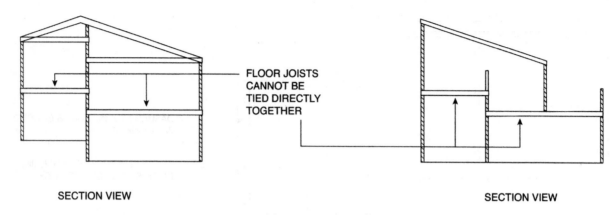

SECTION VIEW   SECTION VIEW

FIGURE 23.14   Portions of floor level offset vertically.

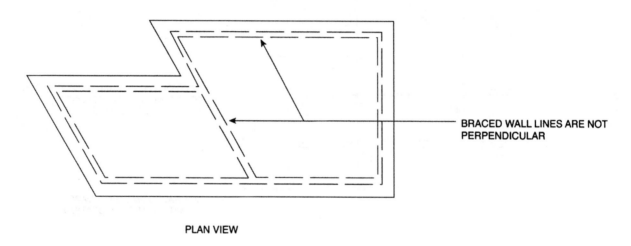

PLAN VIEW

FIGURE 23.15   Braced wall lines not perpendicular.

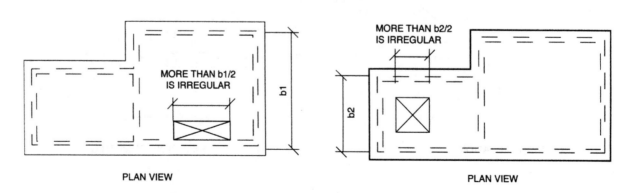

PLAN VIEW   PLAN VIEW

FIGURE 23.16   Opening limitations for floor and roof diaphragms.

MULTIPLE-CHOICE ANSWER KEY

1. B	13. D	25. C	37. C	49. C
2. C	14. D	26. D	38. C	50. B
3. C	15. C	27. A	39. B	51. D
4. B	16. D	28. C	40. B	52. C
5. D	17. D	29. C	41. C	53. B
6. A	18. A	30. A	42. C	54. A
7. B	19. B	31. A	43. A	55. B
8. A	20. C	32. D	44. B	56. B
9. C	21. D	33. C	45. A	57. B
10. D	22. D	34. B	46. B	58. A
11. B	23. A	35. B	47. B	59. C
12. D	24. C	36. C	48. B	60. C

TRUE-FALSE ANSWER KEY

1. T	10. T	19. T	28. F	37. T
2. T	11. T	20. T	29. F	38. F
3. T	12. F	21. F	30. T	39. T
4. T	13. T	22. T	31. T	40. T
5. F	14. T	23. F	32. T	41. T
6. T	15. T	24. T	33. T	42. F
7. F	16. T	25. T	34. T	43. T
8. T	17. F	26. F	35. F	44. T
9. T	18. F	27. T	36. T	45. T

Chapter 24
GLASS AND GLAZING

If glazing has been improperly installed, it can be hazardous to everyone. With changes in the code, it is imperative that you keep up to date. You must make certain that your building structure meets the code requirements. The provisions of Chapter 24 for glass and glazing outline such requirements. This test was designed for you to challenge yourself; use the results to ensure that you are able to meet these requirements.

TRUE-FALSE QUESTIONS

1. The installation of replacement glass is not regulated by the same code requirements used for new installations.

   True   False

2. A decorative composite glazing material made of individual pieces of glass that are embedded in a cast matrix of concrete or epoxy is known as Dalle Glass.

   True   False

3. Each pane of glass is required to bear a manufacturer's mark that designates the type and thickness of the glass or glazing material.

   True   False

4. Tempered glass is exempt from being permanently identified by the manufacturer.

   True   False

5. Tempered spandrel glass is required to be provided with a removable paper marking by the manufacturer.

   True   False

6. Detailed construction documents prepared by registered design professionals are required when one or more sides of any pane of glass are not firmly supported.

   True   False

7. Where interior glazing is installed adjacent to a walking surface, the differential deflection of two adjacent unsupported edges shall not be greater than the thickness of the panels when a force of 50 pounds per linear foot is applied horizontally to one panel at any point up to 42 inches above the walking surface.

   True   False

8. Wired glass with wire exposed on longitudinal edges is the only type of glass allowed to be used with louvered windows or jalousies.

   True   False

9. The effect of wind loads on glass that is sloped 15 degrees or less is a consideration that must be accounted for.

   True   False

10. Code requirements for glass sloped more than 15 degrees from vertical in skylights require specialized design criteria.

   True   False

11. Where used in monolithic glazing systems, heat-strengthened and full-tempered glass shall not have screens installed below the glazing material.

   True   False

12. Screens are not required below any glazing material, including annealed glass, if the walking surface below the glazing material is permanently protected from the risk of falling glass or the area below the glazing material is not a walking surface.

   True   False

13. Skylights set at an angle of less than 45 degrees from the horizontal plane shall be mounted at least 6 inches above the plane of the roof on a curb constructed as required for the frame.

   True   False

14. Skylights shall not be installed in the plane of a roof where the roof pitch is less than 45 degrees from the horizontal.

   True   False

!Codealert

Please note the significant changes for equations in Section 2404.3, 2404.3.3, 2404.3.4, and 2404.3.5 regarding sloped glass that is more than 15 degrees. Section 2404.4 covers designs outside the scope of this section; an analysis or test data for the specific installation must be prepared by a registered design professional.

!Codealert

Glass in elevator enclosures shall be laminated glass conforming to ANSIZ97.1 or 16 CFR Part 1201. Markings as specified in the applicable standard shall be on each separate piece of glass and shall remain visible after installation.

15. Fire department glass access panels shall be made of tempered glass.

    True   False

16. Insulating glass panes, to ensure fire-department access, must not be made of tempered glass.

    True   False

17. Glass used as a wall for a racquetball court must be capable of remaining intact following a test impact.

    True   False

18. Glazing materials shall not be installed in handrails or guards in parking garages, except for pedestrian areas not exposed to impact from vehicles.

    True   False

19. Glass doors shall remain intact following a test impact at the bottom of the door.

    True   False

20. Safety glazing is required in areas identified as hazardous locations.

    True   False

21. The glazing in storm doors must be done with safety glazing.

    True   False

22. Glazing in gymnasiums are subject to human-impact loads.

    True   False

23. Glazing in the glass panels of sliding doors is not required to be made of safety glazing.

    True   False

24. Glass used as a handrail assembly or a guard section shall be constructed of either single full-tempered glass, laminated full-tempered glass, or laminated heat-strengthened glass.

    True   False

25. Glass doors shall remain intact following a test impact at the prescribed height in the center of the door.

    True   False

TRUE-FALSE ANSWER KEY

1. F	6. T	11. F	16. F	21. T
2. F	7. T	12. T	17. T	22. T
3. T	8. F	13. F	18. T	23. F
4. F	9. T	14. T	19. F	24. T
5. T	10. T	15. T	20. T	25. T

Chapter 25

GYPSUM BOARD AND PLASTER

As a contractor, you must be familiar with gypsum board and plaster and how those materials work with the building code. You may be asking yourself why, and I'll tell you. As a contractor, it is important for you to know how the building code is enforced. Chapter 25 offers a general scope and lots of additional information. Without this knowledge, you may neglect the "little" component of the code. While some people discount them, they do exist and are just as important as any other aspect of the code.

MULTIPLE-CHOICE QUESTIONS

1. Weather-exposed surfaces are known as _____.

    a. interior surfaces

    b. exterior surfaces

    c. illegal surfaces

    d. none of the above

2. Cement plaster can be a mixture of _____.

    a. Portland cement

    b. blended cement

    c. masonry cement

    d. any of the above

3. Gypsum board is often referred to as _____.

    a. gypsum wallboard

    b. gypsum sheathing

    c. plaster

    d. both a and b

4. A mixture of calcined gypsum or calcined gypsum, lime, aggregate, and other approved materials is known as _____.

    a. cement plaster

    b. gypsum plaster

    c. veneer plaster

    d. both a and b

5. Gypsum plaster applied to an approved base in one or more coats that normally do not exceed 1/4 inch in thickness is known as:

    a. gypsum veneer plaster

    b. mortar

    c. inadequate

    d. none of the above

6. The installation of gypsum board requires _____.

    a. incandescent light

    b. daylight

    c. eather protection

    d. both a and c

7.   Joint and fastener treatment need not be provided where gypsum board is to receive a decorative finish that would be equivalent to joint treatment. Such decorative finishes include _____.

a. battens

b. wood paneling

c. acoustical finishes

d. all of the above

8.   Joint and fastener treatment need not be provided where gypsum board is used in a _____ system where joints occur over wood framing members.

a. one-layer

b. two-layer

c. multiple-layer

d. none of the above

9.   Joint and fastener treatment need not be provided where gypsum board is _____.

a. square-edge

b. tongue-and-groove edge

c. either a or b

d. none of the above

10.   Gypsum board shall not be used in diaphragm ceilings to resist _____ forces imposed by masonry or concrete construction.

a. seismic

b. lateral

c. special

d. none of the above

11.   Water-resistant gypsum backing board is required as a base for wall tile around _____.

a. bathtubs

b. showers

c. both a and b

d. neither a or b

12.   Water-resistant gypsum backing board is required as a base for wall tile around _____.

a. water-closet compartments

b. showers

c. both a and b

d. neither a or b

13.   The minimum number of coats for gypsum or cement plaster is _____.

a. one

b. two

c. three

d. none of the above

14.   Plaster shall _____ applied directly to fiber insulation board.

a. be

b. not be

c. always be

d. none of the above

15. _____ plaster shall not be applied directly to gypsum lath except under special exceptions.
    a. fiber
    b. cement
    c. wet
    d. none of the above

16. Plaster thickness shall be measured from the _____ and other bases.
    a. face of lath
    b. side of lath
    c. back of lath
    d. none of the above

17. During the preparation and plastering of masonry and concrete, the dash bond coat shall be left undisturbed and shall be moist-cured for not less than _____ hours.
    a. 8
    b. 12
    c. 16
    d. 24

18. Exposed aggregate plaster is permitted to be applied over _____.
    a. concrete
    b. masonry
    c. cement plaster
    d. all of the above

19. When working with an admixture, calcium aluminate cement up to ____ percent of the weight of the Portland cement is permitted to be added to the mix.
    a. 5
    b. 10
    c. 15
    d. 20

20. Gypsum plaster _____ be used on exterior surfaces.
    a. shall not
    b. shall
    c. must
    d. none of the above

TRUE-FALSE QUESTIONS

1. An interior surface is one that is not exposed to weather.
   True   False

2. Horizontal strands of tautened wire attached to surfaces of vertical supports that, when covered with building paper, provide a backing for cement plaster are known as wire backing.
   True   False

3.   Lath and gypsum-board work does not require an inspection.

True   False

4.   Gypsum-board materials and accessories shall be identified by the manufacturer's designation.

True   False

5.   Gypsum wall board is approved for exterior surfaces.

True   False

6.   High-humidity conditions have no noticeable effect on gypsum board.

True   False

7.   Edges and ends of gypsum board shall occur on the framing members, except those that are perpendicular to the framing members.

True   False

8.   Fasteners used on gypsum board must be applied in such a way as not to fracture the face paper with the fastener head.

True   False

9.   Gypsum-board fire-resistance-rated assemblies require all joints and fasteners to be treated except under certain exclusions.

True   False

10.   Gypsum products must conform to ASTM standards.

True   False

11.   Joint and fastener treatment need not be provided when assemblies are tested.

True   False

12.   Joint and fastener treatment need not be provided if gypsum board is to receive a decorative finish that would be equivalent to joint treatment.

True   False

13.   Gypsum board shall not be permitted to be used on wood joists to create a horizontal diaphragm.

True   False

14. End joints of adjacent courses of gypsum board shall not occur on the same joist.
    True   False

15. Gypsum board used in a vertical diaphragm ceiling shall be installed perpendicular to ceiling framing members.
    True   False

16. Water-resistant gypsum backing board shall not be used over a vapor retarder in a shower or bathtub compartment.
    True   False

17. Water-resistant gypsum backing board shall not be used where there will be direct exposure to water.
    True   False

18. Weather protection is required for the storage of materials.
    True   False

19. Gypsum lath is a suitable backing for cement plaster.
    True   False

20. Gypsum sheathing is permitted as a backing for metal or wire fabric lath and cement plaster on walls.
    True   False

21. Wire backing is required under expanded metal lath and paperbacked wire fabric lath.
    True   False

22. Light spraying with water is allowed to keep cement-plaster base coats to an acceptable hardening condition.
    True   False

23. The bedding coat for interior or exterior surfaces shall be composed of one part Portland cement, one part Type S lime, and a maximum of three parts of graded white or natural sand by volume.
    True   False

24. Exterior plaster work may require the installation of weep screeds.
    True   False

25. Plaster coats shall be protected from freezing for a period of not less than 12 hours.

     True   False

MULTIPLE-CHOICE ANSWER KEY

1. B	5. A	9. C	13. C	17. D
2. D	6. C	10. B	14. B	18. D
3. D	7. D	11. C	15. B	19. C
4. B	8. A	12. B	16. A	20. A

TRUE-FALSE ANSWER KEY

1. T	6. F	11. F	16. T	21. F
2. T	7. T	12. T	17. T	22. T
3. F	8. T	13. F	18. T	23. T
4. T	9. T	14. T	19. F	24. T
5. F	10. T	15. F	20. T	25. F

Chapter 26
PLASTIC

This chapter test is designed to strengthen your knowledge about the use of plastic in buildings and new construction. In this test you will possibly find questions from materials to the installation of foam plastics and related items. How well you do depends on how well you've read and studied the chapter. Go ahead with the test and discover how much you really know.

MULTIPLE-CHOICE QUESTIONS

1. Structural plastic panels are those other than _____ that are fastened to structural members, or panels or sheathing and that are used as light-transmitting media in the plane of the roof.

    a. gable vents

    b. ridge vents

    c. skylights

    d. none of the above

2. Reinforced plastic, glass fiber is a plastic reinforced glass fiber having not less than _____ percent of glass fiber by weight.

    a. 5

    b. 10

    c. 15

    d. 20

3. A plastic material that is capable of being changed into a substantially nonreformable product when cured is know as _____.

    a. thermoplastic material

    b. thermosetting material

    c. a light-diffusing system

    d. none of the above

4. Under routine conditions, foam plastic insulation and foam plastic cores of manufactured assemblies shall have a flame spread index of not more than _____.

    a. 25

    b. 35

    c. 50

    d. 75

5. Under routine conditions, foam plastic insulation and foam plastic cores of manufactured assemblies shall have a smoke-developed index of not more than _____.

    a. 250

    b. 350

    c. 450

    d. 575

6. A thermal barrier is not required for foam plastic installed in _____ where the foam plastic insulation is covered on each face by a minimum of one inch thickness of masonry or concrete.

   a. masonry walls

   b. concrete walls

   c. masonry roofs

   d. all of the above

7. A thermal barrier is not required for foam plastic installed in _____ where the foam plastic insulation is covered on each face by a minimum of one inch thickness of masonry or concrete.

   a. masonry walls

   b. concrete floors

   c. masonry roofs

   d. all of the above

8. Foam plastic installed in a cooler or freezer must have a maximum thickness of _____.

   a. 6 inches

   b. 8 inches

   c. 10 inches

   d. 12 inches

9. Exterior doors in buildings of Group R-2 or R-3 that are foam-filled exterior entrance doors to individual dwelling units that do not require a fire-resistance rating shall be faced with _____ or other approved materials.

   a. wood

   b. plastic

   c. metal

   d. none of the above

10. Where pivoted or side-hinged doors are permitted without a fire-protection rating, foam plastic insulation have a flame spread index of _____ or less and a smoke-developed index of not more than 450 shall be permitted as a core material where the door facing is of metal having a minimum thickness of 0.032 inch aluminum or steel having a base metal thickness of not less than 0.016 inch at any point.

    a. 25

    b. 50

    c. 75

    d. 100

11. Foam plastic spray applied to a sill plate and header of Type V construction is subject to the following: The maximum thickness of the foam plastic shall not exceed _____ inches.

    a. 2.5

    b. 3.25

    c. 3.50

    d. 3.75

12. Foam plastic spray applied to a sill plate and header of Type V construction is subject to the following: The density of the foam plastic shall be in the range of _____.

    a. 1.5 to 2.0 pcf   b. super dense

    c. 5 to 15 psi   d. none of the above

13. Foam plastic spray applied to a sill plate and header of Type V construction is subject to the following: The foam plastic shall have a flame spread index of _____ or less.

    a. 15   b. 20

    c. 25   d. none of the above

14. The edge or face of each piece of foam plastic insulation shall bear the _____ of an approved agency.

    a. address   b. phone number

    c. label   d. trademark

15. The edge or face of each piece of foam plastic insulation is required to be marked with _____.

    a. model number   b. serial number

    c. neither a or b   d. both a and b

16. The maximum thickness of plastic interior trim shall not exceed _____.

    a. .25 inch   b. .50 inch

    c. .75 inch   d. 1 inch

17. The maximum width of plastic interior trim shall not exceed _____.

    a. 4 inches   b. 6 inches

    c. 8 inches   d. none of the above

18. The plastic interior trim shall not constitute more than _____ percent of the aggregate wall and ceiling area of any room or space.

    a. 3   b. 5

    c. 7.5   d. 10

19. Plastic veneer shall not be attached to any exterior wall to a height greater than _____ feet above grade.

    a. 16   b. 25

    c. 50   d. 70

20. Sections of plastic veneer shall not exceed _____ square feet in area.

   a. 50   b. 75

   c. 100   d. 300

21. Sections of plastic veneer shall be separated by a minimum of _____ feet vertically when the total allowable square footage requirement is met.

   a. 2   b. 4

   c. 6   d. 10

22. Light-transmitting plastic is required to be marked with identification as to the _____.

   a. density   b. smoke density

   c. material classification   d. all of the above

23. Light-transmitting plastic diffuser shall be supported directly, or indirectly, from ceilings or roof construction by the use of _____.

   a. noncombustible hangers   b. plastic hangers

   c. galvanized wire   d. either a or b

24. Individual panels or units of light-transmitting plastics shall not exceed _____ feet in length.

   a. 5   b. 10

   c. 15   d. 20

25. Individual panels or units of light-transmitting plastics shall not exceed _____ square feet.

   a. 10   b. 15

   c. 25   d. 30

!Codealert

Review section 2603.8 for code requirements in protecting against termites.

TRUE-FALSE QUESTIONS

1. Foam plastic insulation has a density of less than 15 pounds per cubic foot.
    True   False

2. Lenses, panels, grids, and baffles that are part of an electrical fixture are considered to be a light-diffusing system.
    True   False

3. Light-transmitting plastic roof panels are structural plastic panels, other than skylights, that are fastened to structure members, or panels or sheathing and that are used as light-transmitting media in the plane of a roof.
    True   False

4. Plastic glazing is a plastic material that is glazed or set in a frame or sash and not held by mechanical fasteners that pass through the glazing material.
    True   False

5. Foam plastic installed in a maximum thickness of 10 inches in a cooler or freezer shall be protected by an automatic sprinkler system.
    True   False

6. Garage doors using foam plastic insulation complying with Section 2603.3 in detached and attached garages associated with one- and two-family dwellings need to be provided with a thermal barrier.
    True   False

7. Interior trim made of foam plastic in accordance with Section 2604 shall be permitted without a thermal barrier.
    True   False

8. The edge or face of each piece of foam plastic insulation shall bear the label of an approved agency.
    True   False

9. Identifying labels on foam plastic insulation are required to include the manufacturer's model number.
    True   False

10. Foam plastic insulation is not typically allowed for use as an interior wall or ceiling finish in a plenum.

    True   False

11. The presence of termites has no affect on the use of foam plastic insulation.

    True   False

12. The states of Maine, New Hampshire, and Vermont are known to have heavy concentrations of termites.

    True   False

13. The minimum density of the interior trim shall be 25 pcf.

    True   False

14. Light-transmitting plastics shall not be used in bathrooms.

    True   False

15. Light-transmitting plastics shall be permitted in lieu of plain glass in greenhouses.

    True   False

16. Light-transmitting plastics shall not be installed more than 75 feet above grade plane, except as allowed by Section 2607.5.

    True   False

17. Skylights shall be separated from each other by a distance of not less than 6 feet measured in a horizontal plane.

    True   False

18. Light-transmitting plastic interior signs must not exceed 24 square feet in size.

    True   False

19. Light-transmitting plastic interior signs must have their edges and backs fully encased in metal.

    True   False

20. The maximum area of a skylight shall not exceed 50 square feet.

    True   False

21. The aggregate area of light-transmitting plastic interior signs must not exceed 25 percent of the wall area.

     True   False

22. Separation between roof panels is not required in a building equipped throughout with an automatic sprinkler system in accordance with Section 903.3.1.1

     True   False

23. Awnings are not allowed to be constructed of light-transmitting plastics.

     True   False

24. Flat or corrugated light-transmitting plastic skylights shall slope at least four units vertical in 12 units horizontal.

     True   False

25. Dome-shaped skylights shall rise above the mounting flange a minimum distance equal to 10 percent of the maximum span of the dome, but not less than three inches.

     True   False

MULTIPLE-CHOICE ANSWER KEY

1. C	6. D	11. B	16. B	21. B
2. D	7. D	12. A	17. C	22. C
3. B	8. C	13. C	18. D	23. D
4. D	9. A	14. C	19. C	24. B
5. C	10. C	15. D	20. D	25. D

TRUE-FALSE ANSWER KEY

1. F	6. F	11. F	16. T	21. F
2. F	7. T	12. F	17. F	22. T
3. T	8. T	13. F	18. T	23. F
4. T	9. T	14. F	19. T	24. T
5. T	10. T	15. T	20. F	25. T

Chapter 27
ELECTRICAL

National Contractor's Exam Study Guide

This chapter requires no sample questions to be given at this time, since no provisions of the code have been changed for 2006.

Chapter 28
MECHANICAL SYSTEMS

This chapter requires no sample questions to be given at this time, since no provisions of the code have been changed for 2006.

Chapter 29
PLUMBING SYSTEMS

National Contractor's Exam Study Guide

How well did you read this chapter of the code? Are you prepared for this sample test? Do you know and understand the code for replacement systems or equipment? As with all of the sample tests in this study guide, I want you to become aware of what you are not so sure of and build confidence to succeed when taking the real test.

MULTIPLE-CHOICE QUESTIONS

1. A legible sign designating the sex shall be installed in a _____ location near the entrance to each toilet facility.
   a. lighted
   b. readily visible
   c. wall
   d. dry

2. Required plumbing facilities shall be _____.
   a. free of charge
   b. priced at no more than market rate for each use
   c. installed with coin changers
   d. charged for by water usage

3. Employee toilet facilities shall be _____ public toilet facilities.
   a. separate
   b. combined
   c. either a or b
   d. none of the above

4. All single-family residential dwellings are required to have _____.
   a. one toilet
   b. one drinking fountain
   c. one hose bib
   d. none of the above

5. Which of the following types of structures require special consideration for separate plumbing facilities?
   a. stadiums
   b. covered malls
   c. jails
   d. fast-food restaurants

6. The required water closets, lavatories, and showers or bathtubs shall be _____ between the sexes, based on the percentage of each sex anticipated in the occupant load.
   a. of no consequence
   b. distributed equally
   c. both a and b
   d. neither a or b

7.   There are _____ to the rules for separate plumbing facilities.

   a.   no exceptions   b.   exceptions

   c.   complications   d.   no guidelines pertaining

8.   In covered mall buildings, the required public and employee toilet facilities shall not be located more than _____ story(ies) above the space required to be provided with toilet facilities.

   a.   one   b.   two

   c.   three   d.   four

9.   In covered mall buildings, the required public and employee toilet facilities shall not be located more than _____ story(ies) below the space required to be provided with toilet facilities.

   a.   one   b.   two

   c.   three   d.   four

10.   In occupancies other than covered malls, the travel distance to toilet facilities must not exceed _____ feet.

   a.   100   b.   250

   c.   300   d.   500

TRUE-FALSE QUESTIONS

1.   Nightclubs are required to have a minimum of two service sinks.

   True   False

2.   Coliseums are required to have one drinking fountain for every 1000 people rated to attend the coliseums.

   True   False

3.   Places of worship are required to have a minimum of one lavatory for every 200 allowable occupants.

   True   False

4.   Apartment buildings are required to have a minimum of one drinking fountain for each 25 residents allowed to live in the building.

      True   False

5.   Accessible routes to public plumbing facilities may not pass through kitchens.

      True   False

6.   Unisex public restrooms are prohibited.

      True   False

7.   Separate plumbing facilities are not required for single-family dwelling units.

      True   False

8.   Separate plumbing facilities are not required in mercantile occupancies in which the maximum occupant load is 75 or less.

      True   False

9.   Structures and tenant spaces that are intended for public utilization are required to be equipped with public toilet facilities.

      True   False

10.   Separate plumbing facilities are not required in structures or tenant spaces where the total occupant load including employees and customers is 20 or less.

      True   False

MULTIPLE-CHOICE ANSWER KEY

1. B	3. C	5. B	7. B	9. A
2. A	4. A	6. B	8. A	10. D

TRUE-FALSE ANSWER KEY

1. F	3. T	5. T	7. T	9. T
2. T	4. F	6. F	8. F	10. F

Chapter 30

ELEVATORS AND CONVEYING SYSTEMS

If I were to ask you about emergency operations, would you know how to answer? This sample test regarding elevators and conveying systems ask you about these situations. Picking the right choice shouldn't be a guess. And even if you've guessed right, how will those guesses help you with the real test? The real difference in knowing the code and memorizing it is that with memorizing the chances of forgetting the answer are great. See for yourself how well you know and have learned how to apply this code in everyday building circumstances.

MULTIPLE-CHOICE QUESTIONS

1. Hoistways of elevators and dumbwaiters penetrating more than _____ floor(s) are subject to special requirements.

   a. one
   b. two
   c. three
   d. four

2. Hoistways of elevators and dumbwaiters penetrating more than the number of floors specified above are required to be provided with a means of venting _____ to the outer air, in case of fire.

   a. smoke
   b. hot gases
   c. both a and b
   d. none of the above

3. Sidewall-elevator hoistways are _____required to be vented.

   a. sometimes
   b. absolutely
   c. never
   d. always

4. Elevators shall be provided with _____ emergency recall operation and Phase II emergency in-car operation.

   a. phase I
   b. phase II
   c. phase III
   d. phase IV

5. Which of the following is allowed to be installed in the base of an elevator shaft?

   a. floor drains
   b. sumps
   c. both a and b
   d. none of the above

6. Plumbing and mechanical systems _____ be installed in elevator shafts.

   a. may
   b. shall not
   c. must
   d. none of the above

7. Power-operated conveyors, belts, and other material-moving devices shall be equipped with _____ limit switches that will shut off the power in an emergency and stop all operation of the device.

   a. manual   b. automatic

   c. pneumatic   d. pressurized

8. Where standby power is connected to elevators, the machine-room ventilation or air conditioning _____ be connected to the standby power source.

   a. shall not be   b. shall be

   c. can be   d. must never

9. Holes in a machine-room floor for the passage of ropes, cables, or other moving elevator equipment shall be limited so as not to provide greater than ____ inch(es) of clearance on all sides.

   a. 1   b. 2

   c. 3   d. 4

10. Vents shall be located at the top of hoistways and shall open directly to the outer air through _____ ducts.

    a. insulated   b. horizontal

    c. noncombustible   d. vertical

TRUE-FALSE QUESTIONS

1. Rules and regulations are the same for passenger and freight elevators.

   True   False

2. A dumbwaiter is considered a shaft enclosure in terms of complying with the code.

   True   False

3. Door and gate electric contacts and door-operating mechanisms shall be exempt from fire-test requirements.

   True   False

4. The maximum number of elevator cars that are permitted in any single enclosure is three.

   True   False

5.   The maximum number of elevator cars that are permitted in any single enclosure is four.

     True   False

6.   Where two or more elevator cars serve all or the same portion of a building, the elevators shall be located in at least two separate hoistways.

     True   False

7.   Where three or more elevator cars serve all or the same portion of a building, the elevators shall be located in at least two separate hoistways.

     True   False

8.   Where four or more elevator cars serve all or the same portion of a building, the elevators shall be located in at least two separate hoistways.

     True   False

9.   Elevators shall not be in a common shaft enclosure with a stairway.

     True   False

10.   Standby power shall not be manually transferable to all elevators in each bank.

     True   False

11.   When only one elevator is installed, it shall automatically transfer to standby power within 60 seconds after failure of normal power.

     True   False

12.   When only one elevator is installed, it shall automatically transfer to standby power within 5 minutes after failure of normal power.

     True   False

13.   Escalator floor openings shall be enclosed with shafts.

     True   False

14.   Plumbing systems are allowed to be installed in elevator equipment rooms.

     True   False

15.   An approved means of access shall be provided to elevator machine rooms.

     True   False

MULTIPLE-CHOICE ANSWER KEY

1. C	3. C	5. C	7. B	9. B
2. C	4. A	6. B	8. B	10. C

TRUE-FALSE ANSWER KEY

1. F	4. F	7. F	10. F	13. T
2. T	5 T	8. T	11. T	14. F
3. T	6. F	9. T	12. F	15. T

Chapter 31

SPECIAL CONSTRUCTION

Whew, almost there—just a few chapters to go and you'll be finished testing your knowledge of the 2006 International Building Codes. Even though it's been tough, you've hung in there. Throughout this study guide you've learned just how familiar you are with the code. This can only help you in taking the real test. Good luck.

MULTIPLE-CHOICE QUESTIONS

1. A building in which the shape of the structure is maintained by air pressurization of cells or tubes to form a barrel vault over the usable area is known as _____.

    a. a balloon building

    b. not fit for human occupancy

    c. an air-supported structure

    d. none of the above

2. A building in which the shape of the structure is maintained by air pressurization of cells or tubes to form a barrel vault over the usable area is known as _____.

    a. a balloon building

    b. not fit for human occupancy

    c. an air-inflated structure

    d. both a and b

3. A building in which the shape of the structure is maintained by air pressurization of cells or tubes to form a barrel vault over the usable area is known as _____.

    a. a balloon building

    b. an air-supported structure

    c. an air-inflated structure

    d. either b or c

4. An advantage to a double-skin, air-supported structure is _____.

    a. insulation

    b. acoustics

    c. aesthetics

    d. all of the above

5. A nonpressurized structure in which a mast and cable system provides support and tension to a membrane weather barrier and the membrane imparts stability to the structure is known as a(n) _____.

    a. air-inflated structure

    b. air-supported structure

    c. membrane-covered cable structure

    d. none of the above

6. A noncombustible membrane shall be permitted for use as the roof or skylight of any building or atrium of a building of any type of construction, provided that it is at least _____ above any floor, balcony, or gallery.

a. 10 feet

b. 20 feet

c. 22 feet

d. 30 feet

7. A standby-power system for an inflation building system must be able to obtain operating status within _____ seconds of a primary service failure.

a. 20

b. 30

c. 45

d. 60

8. A system capable of supporting the membrane of an inflated structure is required when_____.

a. the structure houses a swimming pool

b. a single-skin structure exists

c. both a and b

d. none of the above

9. A system capable of supporting the membrane of an inflated structure is required when_____.

a. the structure houses a swimming pool

b. occupancy is 50 or more

c. both a and b

d. none of the above

10. Temporary structures require a means of egress with a maximum exit access travel distance of _____ feet.

a. 25

b. 50

c. 75

d. none of the above

11. Temporary structures require a means of egress with a maximum exit-access travel distance of _____ feet.

a. 35

b. 70

c. 85

d. 100

12. With some exceptions, pedestrian walkways and tunnels that are connected shall be considered to be _____.

    a. separate structures   b. illegal

    c. both a and b   d. none of the above

13. Only materials and decorations approved by the _____ shall be located in a pedestrian walkway.

    a. owner   b. contractor

    c. building official   d. either a or c

14. Walkways shall be separated from the interior of the building being served by fire-barrier walls with a fire-resistance rating of not less than _____ hours.

    a. 1   b. 2

    c. 4   d. 6

15. The minimum type of construction of an isolated radio tower not more than _____ feet in height shall be Type IIB.

    a. 25   b. 50

    c. 75   d. 100

16. Radio towers are required to be _____ and effectively grounded.

    a. temporarily   b. permanently

    c. occasionally   d. any of the above

17. When constructing a radio or television tower, the tower must be provided with an adequate foundation, along with anchorage that will resist _____ times the calculated wind load.

    a. two   b. three

    c. four   d. six

!Codealert

See section 3109.5 for entrapment avoidance requirements.

18. Public swimming pools must be enclosed by a fence with a minimum height of _____ feet.

    a. 2   b. 4

    c. 5   d. 6

19. Openings in fences that surround public swimming pools must not permit the passage of a _____-inch-diameter sphere.

    a. 2   b. 4

    c. 6   d. 8

20. When dealing with swimming pools, single- or multiple-pump circulation systems shall be provided with a minimum of _____ suction outlets of an approved type.

    a. two   b. three

    c. four   d. none of the above

TRUE-FALSE QUESTIONS

1. Occupants of air-inflated structures live in the pressurized area used to support the structure.

   True   False

2. When dealing with an air-supported structure, a single-skin design is one in which only the single outer skin is directly against the supporting air pressure.

   True   False

3. There are three basic types of air-supported structures.

   True   False

4. A cable-restrained air-supported structure is another name for a cable-supported structure.

   True   False

5. A nonpressurized building wherein the structure is composed of a rigid framework to support a tensioned membrane that provides a weather barrier is known as a membrane-covered frame structure.

   True   False

6. Membrane structures must not exceed one story in height.

    True   False

7. Air-supported and air-inflated structures must be provided with primary and auxiliary inflation systems to meet code requirements.

    True   False

8. Inflation systems are required to be designed to prevent over pressurization of the system.

    True   False

9. Blowers for inflation systems are prohibited from having backdraft check dampers.

    True   False

10. Standby power for an inflation system must be capable of operating independently for a minimum of eight hours.

    True   False

11. A building permit is required for a temporary structure that covers an area in excess of 120 square feet.

    True   False

12. Temporary structures erected to house more than six people and cover an area of 100 square feet require a building permit.

    True   False

13. Pedestrian walkways must be made from noncombustible construction.

    True   False

14. Combustible construction shall be permitted where connected buildings are of combustible construction.

    True   False

15. Separation between a tunneled walkway and the building to which it is connected shall not have less than one-hour fire-resistant construction.

    True   False

16. A retractable awning is a cover with a frame that retracts against a building or other structure to which it is entirely supported.

    True   False

17. Signs are required to be designed, constructed, and maintained in accordance with code requirements.

    True   False

18. Radio and television towers must not be constructed with step bolts and ladders.

    True   False

19. Guy wires must not cross or encroach upon any street.

    True   False

20. Radio and television towers are required to be constructed of approved corrosion-resistant noncombustible material.

    True   False

21. Television towers must be designed for the dead load plus the ice load in regions where ice formation occurs.

    True   False

22. To be considered a swimming pool, the pool must have a minimum depth of 12 inches.

    True   False

23. A hot tub could be considered to be a swimming pool.

    True   False

24. Barriers for swimming pools are to be located so as to prohibit permanent structures from being used to climb the barriers.

    True   False

25. Suction outlets for swimming pools are to be designed to produce circulation throughout the pool.

    True   False

MULTIPLE-CHOICE ANSWER KEY

1. D	5. C	9. C	13. C	17. A
2. C	6. B	10. D	14. B	18. B
3. C	7. D	11. D	15. D	19. B
4. D	8. A	12. A	16. B	20. A

TRUE-FALSE ANSWER KEY

1. F	6. T	11. T	16. T	21. T
2. T	7. T	12. F	17. T	22. F
3. F	8. T	13. T	18. F	23. T
4. F	9. F	14. T	19. T	24. T
5. T	10. F	15. F	20. T	25. T

Chapter 32

ENCROACHMENTS INTO THE PUBLIC RIGHT-OF-WAY

National Contractor's Exam Study Guide

Congratulations on making it to this point in the 2006 International Building Code study guide. You must be proud of yourself; I know I am. Only two more sample tests after this one, and you will know that when you are ready to take the real test, you have the confidence and security of understanding the code.

TRUE-FALSE QUESTIONS

1.   Drainage water collected from a roof or awning must not flow over a public walking surface.
       True   False

2.   The code allows water collected from a marquee or a condensate from mechanical equipment to flow over a public walking surface.
       True   False

3.   Areaways shall be protected by grates, guards, or other means.
       True   False

4.   Steps must not extend more than 10 inches.
       True   False

5.   Steps must not extend more than 12 inches.
       True   False

6.   Encroachments that extend 15 feet or more above grade shall be limited.
       True   False

7.   Steps involved with encroachments are required to be guarded by approved devices that are not less than 3 feet high.
       True   False

8.   Columns or pilasters, including bases and molding, shall not project more than 12 inches.
       True   False

9.   Lintels shall not project more than 12 inches.
       True   False

10. The vertical clearance from a public right-of-way to the lowest part of any awning, including a valance, shall be a minimum of 10 feet.

True   False

11. The vertical clearance from a public right-of-way to the lowest part of any awning, including a valance, shall be a minimum of 8 feet.

True   False

12. The vertical clearance from a public right-of-way to the lowest part of any awning, including a valance, shall be a minimum of 7 feet.

True   False

13. The vertical clearance from a public right-of-way to the lowest part of a pedestrian walkway shall be a minimum of 10 feet.

True   False

14. The vertical clearance from a public right-of-way to the lowest part of a pedestrian walkway shall be a minimum of 15 feet.

True   False

15. Stanchions or columns that support awnings shall be located not less than 2 feet in from the curb line.

True   False

TRUE-FALSE ANSWER KEY

1. T	4. F	7. T	10. F	13. F
2. F	5. T	8. T	11. F	14. T
3. T	6. F	9. F	12. T	15. T

Chapter 33

SAFEGUARDS DURING CONSTRUCTION

National Contractor's Exam Study Guide

This chapter explains the code provisions for safeguards during construction. Many people are now building homes, either hiring contractors or doing it themselves. Everywhere you turn, new businesses are being built as well.

Although it is a short chapter, the provisions of Chapter 23 are not to be taken lightly. From remodeling to automatic sprinkler systems, safety during construction must be taken seriously. Construction is all about multitasking. Make sure you practice multitasking every day while following the code to safeguard the public, the workers, and any adjoining property during construction. This is also true for any equipment that you are using at the building site. The bottom line is to read and learn the chapter so you can apply the provisions of this code and any other codes that may coexist.

The results of this practice test will show you the areas that you know and perhaps other areas that you need to brush up on. Good luck on the test, and we'll meet again in the next and last chapter!

MULTIPLE-CHOICE QUESTIONS

1. Any building that is undergoing renovation is required to have maintained sanitary conditions, unless _____.

   a. the building is of a commercial nature

   b. the building is unoccupied

   c. the building is habitable

   d. none of the above

2. Any building that is undergoing renovation is required to have maintained sanitary conditions, unless _____.

   a. the building is of a commercial nature

   b. the building is one story tall

   c. the building is habitable

   d. none of the above

3. Any building that is undergoing renovation is required to have fire-protection equipment, unless _____.

   a. the building is of a commercial nature

   b. the building is unoccupied

   c. the building is habitable

   d. none of the above

4.   Waste materials must be removed in a manner that will not injure or damage_____.

   a. people   b. adjoining property

   c. public right-of-ways   d. all of the above

5.   When performing site work, tree stumps and roots that are within _____ inches of the ground surface to house a building must be removed.

   a. 6   b. 12

   c. 16   d. 18

6.   Cut slopes for permanent excavations shall not be steeper than one unit vertical in ____ units horizontal.

   a. two   b. three

   c. four   d. none of the above

7.   Pedestrian walkways must have a minimum width of _____.

   a. 30 inches   b. 36 inches

   c. 42 inches   d. 48 inches

8.   With a building that has a height of no more than 8 feet and that is less than 5 feet from the construction to the lot line, contractors must provide _____ as protection for pedestrians.

   a. none   b. railings

   c. barriers   d. none of the above

9.   With a building that has a height of no more than 8 feet and that is less than 5 feet from the construction to the lot line, but exceeding one-half the height of construction, contractors must provide _____ as protection for pedestrians.

   a. none   b. railings

   c. barriers   d. none of the above

10. With a building that has a height of no more than 8 feet and that is less than 5 feet from the construction to the lot line, but exceeding one-fourth the height of construction, contractors must provide _____ as protection for pedestrians.

   a. covered walkways   b. railings

   c. barriers   d. both a and c

11. Barriers shall be a minimum of 8 feet in height and shall be placed on the side of the walk-way _____ the construction.

    a. nearest   b. furthest from

    c. to the right of   d. to the left of

12. When a building is constructed with a height of more than 50 feet, at least _____ shall be provided unless one or more of the permanent stairways are erected as the construction progresses.

    a. one temporary stairway   b. two temporary stairways

    c. two fire extinguishers   d. none of the above

13. Every excavation on a site located 5 feet or less from the street lot line shall be enclosed with a barrier that is not less than _____ feet in height.

    a. 3   b. 4

    c. 5   d. 6

14. Adjoining _____ property shall be protected from damage during construction, remodeling, and demolition work.

    a. private   b. public

    c. both a and b   d. none of the above

15. Any person making or causing an excavation shall provide written notice to the owners of adjoining buildings advising them that the excavation is to be made and that the adjoining buildings should be protected. The written notification must be delivered not less than _____ days prior to the scheduled starting date of the excavation.

    a. 5   b. 10

    c. 14   d. 30

!**Code**alert

Temporary stairway-floor number signs shall be provided in accordance with the requirements of Section 1020.1.6.

16. All structures under construction, alteration, or demolition shall be provided with not less than _____.

    a. one fire extinguisher   b. two fire extinguishers

    c. one sprinkler system   d. none of the above

17. In buildings where an automatic sprinkler system is required by code, it shall be unlawful to occupy any portion of a building or structure until the automatic sprinkler-system installation has been _____.

    a. tested   b. approved

    c. both a and b   d. none of the above

18. Operation of sprinkler control valves shall be permitted only by properly _____ personnel and shall be accompanied by notification of duly designated parties.

    a. authorized   b. trained

    c. intended   d. none of the above

19. Required means of egress shall be maintained _____ during construction.

    a. during working hours   b. at all times

    c. during daytime hours   d. after dark

20. Waste materials must be removed in a manner that prevents injury or damage to:

    a. surrounding land   b. public right-of-ways

    c. pedestrians   d. all of the above

TRUE-FALSE QUESTIONS

1. Construction equipment must be stored and placed so as not to endanger the public.

   True   False

2. With two exceptions, required exits must be maintained at all times during remodeling.

   True   False

3. When an existing building is unoccupied and being remodeled, fire-protection devices must be maintained at all times.

   True   False

4. Waste materials must be removed in a manner that prevents injury of persons.

True   False

5. Pedestrian protection is a key element in the demo work of a building.

True   False

6. Destruction of a party-wall balcony or horizontal exit is not allowed.

True   False

7. Vacant lots where buildings have been demolished are not regulated by the building code.

True   False

8. During demolition work, provision must be made to prevent the accumulation of water or damage to any foundations on the premises or adjoining property.

True   False

9. Slopes for permanent fill shall not be steeper than one unit vertical in two units horizontal.

True   False

10. Before completion, loose or casual wood shall be removed from direct contact with the ground under a building.

True   False

11. Signs are required to direct pedestrian traffic during construction and remodeling.

True   False

12. Construction railings must have a minimum height of 46 inches.

True   False

13. Barriers designed for the protection of pedestrians must have a minimum height of 8 feet.

True   False

14. Covered walkways for pedestrians shall in no case be allowed to have a live load of less than 200 pounds per square foot.

True   False

15. Every excavation on a site located 5 feet or less from the street lot line shall be enclosed with a barrier not less than 6 feet high.

True   False

16.   Adjoining public property shall be protected from damage during construction, remodeling, and demolition work.

    True   False

17.   Adjoining private property shall be protected from damage during construction, remodeling, and demolition work.

    True   False

18.   Buildings with three or more stories in height are required to be equipped with standpipes.

    True   False

19.   A water supply for fire protection, either temporary or permanent, must be made available as soon as combustible material accumulates.

    True   False

20.   Required means of egress shall be maintained at all times during demolition.

    True   False

21.   Pedestrian protection must be provided prior to the demolition of a building.

    True   False

22.   No fill or other surcharge loads are allowed adjacent to any building or structure.

    True   False

23.   Covered walkways are required to have adequate lighting.

    True   False

24.   Barrier design requires that top and bottom plates be made of 2-inch-x-4-inch lumber.

    True   False

25.   Fire extinguishers are required for all structures under construction, alteration, or demolition.

    True   False

MULTIPLE-CHOICE ANSWER KEY

1. B	5. B	9. A	13. D	17. C
2. D	6. A	10. D	14. C	18. A
3. B	7. D	11. A	15. B	19. B
4. D	8. B	12. A	16. A	20. D

TRUE-FALSE ANSWER KEY

1. T	6. F	11. T	16. T	21. T
2. T	7. F	12. F	17. T	22. F
3. T	8. T	13. T	18. F	23. T
4. T	9. T	14. F	19. T	24. T
5. T	10. T	15. T	20. T	25. T

Chapter 34

EXISTING STRUCTURES

It is not difficult to figure out what this chapter of the code governs. Yes, there are codes in place for these types of structures. Remember that with this chapter and others there may be additional codes that you must comply with as well. The building official can and will check. What the building official giveth, the building official can take away. And until a second inspection results in a "thumbs up," the inspector will put a stop-work order on your project. Think of the loss of time and money should you choose to disobey the code. The codes for existing structures include provisions for remodeling, repair, and alteration. Take this sample test to see which if any areas you should brush up on before taking the real test. Good luck.

MULTIPLE-CHOICE QUESTIONS

1. Areas that contain a primary function include but are not limited to _____.
    a. the dining area of a cafeteria
    b. a janitorial closet
    c. a mechanical room
    d. none of the above

2. Areas that contain a primary function include but are not limited to _____.
    a. the customer-service lobby of a bank
    b. a janitorial closet
    c. a locker room
    d. none of the above

3. Areas that contain a primary function include but are not limited to _____.
    a. boiler rooms
    b. janitorial closets
    c. mechanical rooms
    d. none of the above

!Codealert

A primary function is a major activity for which the facility is intended.

4.   Areas that contain a primary function include but are not limited to _____.

   a. supply storage rooms   b. janitorial closets

   c. mechanical rooms   d. none of the above

5.   Areas that contain a primary function include but are not limited to _____.

   a. the dining area of a cafeteria

   b. a janitorial closet

   c. meeting rooms in a conference center

   d. both a and c

6.   Areas that contain a primary function include but are not limited to _____.

   a. restrooms   b. offices

   c. mechanical rooms   d. none of the above

7.   When altering or making an addition on a building that is located in a flood-hazard area, the code for new construction in flood-hazard areas must be observed when the changes are _____.

   a. cosmetic   b. substantial

   c. both a and b   d. none of the above

8.   Additions or alterations to an existing structure shall not increase the force in any structural element by more than _____ percent, unless the increased force on the element is still in compliance with the code for new structures, nor shall the strength of any structural element be decreased to less than that required by the code for new structures.

   a. 3   b. 5

   c. 8   d. 10

9.   An addition must not increase the seismic forces in any structural element of an existing structure by more than _____ percent cumulative since the original construction, unless the element has the capacity to resist the increased forces as determined in accordance with ASCE7.

   a. 3   b. 5

   c. 8   d. 10

10. When dealing with fire escapes, the stairs must have a minimum width of _____ inches.

   a. 12   b. 20

   c. 22   d. 24

> **!Code**alert
>
> Additions, alterations, or modification or change of occupancy of existing buildings shall be in accordance with this section for the purposes of seismic considerations.

11. Fire escapes shall comply with the code and shall not constitute more than _____ percent of the required number of exits.

    a. 15   b. 25
    c. 33   d. 50

12. For fire escapes that are located on the front of a building and that project beyond the building line, the lowest landing shall not be less than ____ feet.

    a. 3   b. 5
    c. 7   d. 10

13. For fire escapes that are located on the front of a building and that project beyond the building line, the lowest landing shall not be more than ____ feet.

    a. 5   b. 7
    c. 12   d. 15

14. Fire escapes are required to be designed to support a live load of _____ pounds per square inch.

    a. 25   b. 35
    c. 100   d. 125

15. Fire-escape risers are to be no more than, and treads not less than, _____ inches.

    a. 5   b. 6
    c. 8   d. none of the above

16. Doors and windows along a fire escape shall be protected with _____ opening protection.

    a. 1/4 hour   b. 1/2 hour

    c. 3/4 hour   d. 1 hour

17. A certificate of occupancy shall be issued when it has been determined that the requirements for the new occupancy classification have been _____.

    a. established   b. met

    c. neither a or b   d. both a and b

18. Where it is technically infeasible to alter performance areas to be on an accessible route, at least ____ of each type of performance area shall be made accessible.

    a. one   b. two

    c. three   d. none of above

19. Accessible ramps that are steeper than 1:10 but not steeper than 1:8 shall have a maximum rise of _____.

    a. 2 inches   b. 3 inches

    c. 5 inches   d. 6 inches

20. Accessible ramps that are steeper than 1:12 but not steeper than 1:10 shall have a maximum rise of _____.

    a. 4 inches   b. 6 inches

    c. 8 inches   d. 10 inches

> **!Code**alert
>
> When a change of occupancy results in a structure being re-classified to a higher occupancy category, the structure shall conform to the seismic requirements for a new structure.

TRUE-FALSE QUESTIONS

1. A primary function is a major activity for which the facility is intended.
   True   False

2. Areas that contain a primary function include but are not limited to the customer-service lobby of a bank.
   True   False

3. Restrooms contain a primary function.
   True   False

4. Corridors contain a primary function.
   True   False

5. Mechanical rooms do not contain a primary function.
   True   False

6. When work in an existing structure is considered to be technically unfeasible, the work is not likely to be done.
   True   False

7. Additions to any building or structure shall comply with the requirements of the code for new construction.
   True   False

8. Alterations to any building or structure shall comply with the requirements of the code for new construction.
   True   False

9. New additions to existing structures in flood-hazard areas are allowed under a grandfather clause.
   True   False

10. Seismic considerations do not apply to additions to existing structures.
    True   False

11. An addition that is structurally independent from an existing structure shall be designed and constructed according to the seismic requirements for new structures.

True   False

12. Fire escapes shall not constitute any part of the required means of egress in new buildings.

True   False

13. Nonstructural alterations or repairs to an existing building are permitted only when they are approved by the local plumbing inspector.

True   False

14. Existing fire escapes shall continue to be accepted as a component in the means of egress in existing buildings only.

True   False

15. New fire escapes are allowed to incorporate ladders.

True   False

16. New fire escapes are not allowed to incorporate access by windows.

True   False

17. New fire escapes for existing buildings shall be permitted only where exterior stairs cannot be utilized due to lot lines limiting stair size or to sidewalks.

True   False

18. Glass replacements are not regulated by the code.

True   False

!Codealert

For existing buildings located in flood hazard areas established in Section 1612.3, if the alterations and repairs constitute substantial improvement of the existing building, the existing building shall be brought into compliance with the requirements for new construction for flood design.

19. Any structure that is moved into or within a jurisdiction must comply with the code provisions established for new construction.

     True   False

20. When a change of occupancy results in the reclassification of a structure to a higher occupancy category, the structure shall conform to the seismic requirements for a new structure.

     True   False

21. At least two accessible routes from a site arrival point to an accessible entrance shall be provided.

     True   False

22. When altering jury boxes and witness stands, the defined area must be equipped for wheelchair access.

     True   False

23. Thresholds for doors are required to have beveled edges on each side.

     True   False

24. Included within the means-of -egress category are the configuration, characteristics, and support features for means of egress in a facility.

     True   False

25. Historic buildings require no additional consideration by the code.

     True   False

MULTIPLE-CHOICE ANSWER KEY

1. A	5. A	9. D	13. C	17. B
2. A	6. B	10. C	14. C	18. A
3. D	7. B	11. D	15. C	19. B
4. D	8. B	12. C	16. C	20. B

TRUE-FALSE ANSWER KEY

1. T	6. T	11. T	16. T	21. F
2. T	7. T	12. T	17. T	22. F
3. F	8. T	13. F	18. F	23. T
4. F	9. F	14. T	19. T	24. T
5. T	10. F	15. F	20. T	25. F